William Hammond Hall

Outline of Matter and Advance Sheets of the Report on the Legislative, Administrative, Technical, and Practical Problems of Irrigation

William Hammond Hall

Outline of Matter and Advance Sheets of the Report on the Legislative, Administrative, Technical, and Practical Problems of Irrigation

ISBN/EAN: 9783744644648

Printed in Europe, USA, Canada, Australia, Japan

Cover: Foto ©Suzi / pixelio.de

More available books at **www.hansebooks.com**

OUTLINE OF MATTER, AND ADVANCE SHEETS

OF THE

REPORT

ON THE

Legislative, Administrative, Technical, and Practical

PROBLEMS OF IRRIGATION,

IN COURSE OF PREPARATION AND PUBLICATION.

WM. HAM. HALL,
State Engineer.

SACRAMENTO:
STATE OFFICE........ JAMES J. AYERS, SUPT. STATE PRINTING.
1884.

NOTICE.

The 304 pages of matter within these covers have been thus brought together and bound for transmission to the Legislature as an exhibit of the extent and character of the REPORT ON THE PROBLEMS OF IRRIGATION, now in course of preparation and publication by the State Engineer.

The first 32 pages contain a *Table of Contents*, or *Outline of Matter*, for each chapter of the entire work, from which a fair idea may be formed of its scope, system of arrangement, degree of completeness, and general character.

The 272 pages thereafter—32 to 304—contain the text of about three fifths of the First Book of the work—namely, the papers on the IRRIGATION LEGISLATION AND ADMINISTRATION of the *Romans*, of the *French*, and of the *Italians*—from which a fair idea may be formed of the character of the matter and its treatment in detail.

As planned, the work consists of Three Parts, made up of Seven Books, with subject-matter as indicated by the following titles:

PART I.

The Social, Political, and Legal Problems of Irrigation.

BOOK I—The Laws of Waters and Water-courses, and the Customs, Laws, and Policies with respect to Irrigation, in Civil Law countries.

BOOK II—The Laws of Waters and Water-courses, so far as these directly affect irrigation questions, and the Customs, Laws, and Policies, with respect to Irrigation, in Common Law countries.

PART II.

The Physical, Practical, and Technical Problems of Irrigation.

BOOK III—The physical questions of Water supply, Conservation, and Division for purposes of Irrigation in California.

BOOK IV—The existing Works, Practice, and Possibilities of Irrigation in California.

BOOK V—The technical questions of Water Distribution and Use in the practice of Irrigation in, and as applied to, California.

PART III.

The Planning, Construction, Operation, and Maintenance of Irrigation Works.

BOOK VI—Of Works for the Interception and Storage of Waters for Irrigation.

BOOK VII—Of Works for the Diversion, Conducting, and Applying Waters in Irrigation.

The first five Books of the work, with their necessary appendices, will make up four volumes, each of 450 to 550 pages, of the size and style of the sheets now printed, varying with the provision which may be made for printing, illustration, revising, and editing the matter now available.

The matter for the last two Books is being collected incidentally to the preparation of the first five. It may be published within the compass of one volume or may be extended to two volumes, according to the provision made for completion and the operations of publication as above.

For further information concerning this report, see Report of the State Engineer to the Legislature, for the two years ending with December 31, 1884.

BOOK I.

IRRIGATION LEGISLATION AND ADMINISTRATION IN COUNTRIES UNDER THE CIVIL LAW.

A.—The Roman Empire—Introductory.

CHAPTER I.—*The Roman Laws and Administrative Policy with respect to Waters and Water-courses.*

INTRODUCTION—IMPORTANCE OF THE ROMAN LAWS OF WATERS:—Time and circumstances of forming; The fountain head of legal reasoning; The basis of modern Civil law.

SECTION I—RIGHT OF PROPERTY IN WATERS AND WATER-COURSES:—Property classified; Common property defined; Running waters, common property; Public property defined; Navigable rivers and important streams, public property; Unimportant streams, private property; Rivers defined; River banks and beds; Ownership and use of banks and beds; Resumé as to ownership.

SECTION II—CONTROL OF PUBLIC RIVERS AND WATERS:—The rights of navigation and fishery; Guarding of channels, banks, and beds; Unlicensed works prohibited; Diversion prohibited; Construction and maintenance of works; Riparian right to protect banks; Diversion of public waters; Public waters always public; Appropriation not allowed; The waters devoted to public use; Permits granted to divert and use; Use of public waters; Terms of grants or permits; Exercise of water privileges; Public springs; Public reservoirs; Prescriptive rights; Waste prohibited.

SECTION III—CONTROL OF WATERS IN PRIVATE WORKS:—Springs on private lands, private property; Rights to use spring waters; Prescription; Agreement; Spring waters become common property; Riparian right to use; Water in private works, private property; Use of private waters; Waste prohibited.

SECTION IV—THE RIGHT OF WAY TO CONDUCT WATERS:—Servitudes; Dominant estates; Servient estates; Servitudes classified; The servitude to conduct water; Acquirement of right; Prescription; Agreement; Condemnation of right of way for public works; Servitude of right of way for private works; Permission to conduct water across public property; The servitude to draw water.

B.—French Irrigation Legislation and Administration.

CHAPTER II—FRANCE ([1]); *The Right of Property in and Control of Water-courses in France.*

SECTION I—ORIGIN OF PROPERTY RIGHTS AND OWNERSHIP OF STREAMS:—*Basis of property rights;* Downfall of Roman rights; Merovingian system; Feudal tenure; Downfall of Feudalism; The modern monarchy.

Ownership and control of navigable streams; Navigable and floatable streams, public property; Possible arbitrary application of the rule; The Code Napoleon; Floatable streams, public property; The edict of Moulines; Inalienability of the public domain.

Ownership and control of streams not navigable nor floatable; Riparian claims of ownership; Water-courses not navigable nor floatable, common property; Riparian claims to ownership of stream beds; Government control of channels.

Riparian claims to waters as a property; Riparian claims to waters as a common property ¡ General claim of waters as a common property of all the people; Riparian right to use waters, as a servitude; The waters are a common property of all the people; The beds a common property so long as covered.

SECTION II—WATER LAWS AND REGULATIONS:—*Moving causes of development;* Agriculture and irrigation; Manufacturing and water power; Internal transportation made necessary; Inundations; Sanitary necessities.

Special regard for irrigation; Agriculture a leading interest; Irrigation ranked in the laws, above other uses except domestic use and navigation; The administration favors irrigation; Liberality of French water laws.

Classification of water laws; The earliest laws; Statutory law; Two branches of a sort of Common law; Judicial decisions; Administrative rulings; No one, general, water-law or code in France.

SECTION III—THE ADMINISTRATION:—*Antiquity of French supervision of public and common property;* The executive branch of government; Its decrees, instructions, regulations, etc.; Administration of non-navigable streams; Administration of navigable streams.

Administrative purpose and policy; Regulation of works and waters on floatable streams; Regulation of works in streams not floatable.

Government organization; France; Government; Legislative branch; Executive branch; Council of state; Ministry; Departmental governments; Prefects and councils; Arrondissements; Sub-prefects; Communes; Mayors and municipal councils.

The administrative system; Line of administrative duty; Extent of a department and of a commune.

The bureau of public works; Minister of public works; Advice of the council of state; Prefects' executive duties.

The engineering department; Education of the engineers; Their field of duty; The conductors or superintendents; Their preparation and duty.

Administrative working; Powers and duties of prefects; Powers and duties of engineers.

Navigation and river guards; River regulations; Policing of streams; River guards: Duties and compensation; Necessity for river guards.

CHAPTER III—FRANCE (²); *Water Privileges and the Administration of Navigable and Floatable Streams.*

SECTION I—WATER PRIVILEGES:—*The uses to which water is put and the regulation of its use;* Irrigation, manufacturing, industrial works, and municipal uses; Government regulation of its use.

The object of administration; Interference, not the object; Promotion of harmony and prevention of conflict, the object; Worthiness of and necessity for the principle.

Rivers and river works in France; The necessity for rivers conservancy; The systems on the lower and on the higher rivers.

Navigable and non-navigable rivers; Public rivers defined; Non-navigable tributaries; Non-navigable arms of navigable rivers.

Forms of organization of irrigation enterprise; Private individual enterprise; Associate or coöperative enterprise; Speculative corporate companies.

Applications and formalities; For water privileges on non-floatable streams; Application, preliminary inquiry, publication, engineers' reports, final inquiry; Water privilege grants on public streams; Applications, reports, plans, projects; Inquiries, engineers' reports, ministers' decisions.

Water-right grants; The case of the Bourne canal; Obligations of the grantees; Conditions of the concession; Privileges of the grantees; Benefits to the company.

SECTION II—REGULATION OF WORKS:—*Government improvement of navigable rivers;* Canalization of upper rivers; Movable dams; The hydraulic service of the public works bureau; Its extent, field, and duty.

Organization for agricultural hydraulic works; The principles of coöperation and compulsion; Cases when water is an enemy and when an auxiliary.

Regulation of the construction of works; Regulation of the construction and height of dams; Combination of navigation with water-power or irrigation dams; payment for works; Regulation of the construction of headworks; Payment for works; Sluiceways; Grades.

SECTION III—OPERATION AND MAINTENANCE: *General maintenance of works;* Of works of navigation; Of joint navigation and irrigation or water-power works; Of private works; State, joint, and private expense.

Cleaning and dredging of channels; Public expense; Contributions of dam owners and water employers; Contributions of riparian proprietors.

Police of streams; Violations of laws; Severe penalties; Laws of ancient dates; Penalties modified; Powers of councils of prefecture; Compulsory removal of objectionable works; Administrative duties of prefects.

Water privilege rents: All privileges subject to rent charges; Rates for water-power; Rates for irrigation; Rates for industrial uses; Rates for municipal uses; Nominal rates; Duties of engineers; Collection of back rents; Revision of rents every thirty years; Exemption of rights antedating 1566.

CHAPTER IV—FRANCE (³); *Water Rights on, and the Administration of Non-navigable Streams.*

SECTION I—RIGHTS TO THE USE OF WATER:—*Water-rights previous to the time of the Code Napoleon:* Riparian claims to absolute control of streams and waters; Conflicting interests; The administrative view; The waters a common property of all people.

Riparian water-rights under the code; Claim of exclusive right; Claim under article 552; Province and duty of the courts; Decisions upholding the administrative view; Attitude of the administration under the decisions.

The riparian water-right, and the right of way; Previous to the law of 1845; The right of way law of 1845; The dam privilege law of 1847; Control of the fall or slope of a stream; Decision of the court of cassation; Backing up of water permitted; Exclusive riparian servitude on use of waters.

The nature of the riparian right to water; No element of ownership; No semblance of the principle of prior appropriation; Merely an undefined and unsegregated part of a common right; Subject to regulation.

The right of irrigation; Absorption of water; Drainage and residue to be returned; Each case one for equitable administration.

SECTION II—SUPERVISION OF CONSTRUCTION OF WORKS:—*The decentralization of the administration;* More power vested in the departmental authorities, the prefects, and engineers; Extent of their powers and duties.

Nature of the powers held by prefects; Police measures compel respect for public interests; Regulation of division of waters; Regulation of construction of works.

Applications for sanctions to construct works; Right to water must be established; Formalities, publication, inquiries, engineering reports; Conditions attached to permits.

Dams and Headworks; Legal heights of dams; Determination and marking; Sluiceways and weirs; Movable dams for irrigation; Dimensions and form of outlets.

SECTION III—REGULATION AND OPERATION—WORKS AND WATERS:—*The necessity for regulation and administration;* Individual unreasonableness; Conflicting decisions; Varying physical conditions; Recognition of the necessity for regulation.

Administrative authority to make regulations; Origin of the authority found in necessity and gradual development; The promotion of harmony and prevention of abuse.

The principles adhered to; The points to be met; The method of meeting them; General rules as to division of waters; Division by measurement and by turns; Ancient custom governing; Decree of July, 1872.

Policing of Water-courses; The formula prescribed in 1878 by ministerial circular.

CHAPTER V—FRANCE (⁴); *Rights of Property in Springs, and Rights to the Use of Spring Waters.*

SECTION I—OWNERSHIP AND CONTROL OF SPRINGS:—*Absolute ownership;* Ownership of the land carries with it ownership of a spring on it; This doctrine for a long time strongly opposed; But it has been upheld; But the right to use spring waters may be lost.

SECTION II—ACQUIRED RIGHTS TO SPRING WATERS:—*Public and private use of springs;* The necessities of communities; The interests of navigation; Private acquirement of right by title and by prescription; Servitudes, resulting from prescriptive use and from divisions of estates.

SECTION III—DRAINAGE AND OTHER RIGHTS:—*Natural right of drainage;* Restrictions on the extension of the right.

The right to dig or bore for water; Extent of the privilege; Forfeiture.

CHAPTER VI—FRANCE (⁵); *The Right of Way to conduct Water, and the Right to abut a Dam.*

SECTION I—RIGHTS FOR WORKS OF PUBLIC IMPORTANCE:—*Condemnation for works of public utility;* The laws of 1836 and of 1841; Administrative inquiry; Special laws of authorization; Way for main and secondary works; Embarrassment previous to right-of-way law of 1845.

SECTION II—RIGHTS FOR PRIVATE WATER WAYS:—*The servitude of right-of-way;* Opposition to its establishment; the law of 1845.

The servitude of right-to-abut-a-dam; The complement of the former servitude; The law of 1847; The application of these laws.

CHAPTER VII—FRANCE (⁶); *Irrigation Enterprise and Organization.*

SECTION I—GOVERNING INFLUENCES:—Climatic and social influences; Irrigation not generally appreciated; Small landholdings and jealousy of rights; Poverty of peasantry and indifference of capital; High valuation of lands; Heavy cost of works; Riparian rights question.

SECTION II—IRRIGATION COMPANIES AND ASSOCIATIONS:—*Forms of association;* Speculative companies; Association of landholders; Necessity and advantage of association; Causes which retarded appreciation.

Syndicate associations; An analysis of the law of 1865; Free syndicate associations; Authorized syndicate associations; Prefectorial power; Governmental policy; The principles of coöperation and compulsion.

CHAPTER VIII—FRANCE (⁷); *Governmental Policy and Irrigation Concessions.*

SECTION I—FEATURES OF POLICY AND FORMS OF ENTERPRISE: *Political and social conditions;* Not such as to warrant irrigation being made a general national work;· Contrast between the case of France and those of India and Egypt.

Forms of governmental encouragement; Tax rebate on advanced values due to irrigation; Loans, advances, subsidies, guarantees; Main works built for associations; Main works built for state management; Premiums on irrigation examples; Collection and publication of irrigation statistics; Statistical atlas of irrigation.

SECTION II—NOTABLE INSTANCES OF ENTERPRISE AND ENCOURAGEMENT:—The Canal des Alpines, Canal Carpentras, Canal of Cadenet, Canal of St. Martery, Canal of Siagne, Canal of Siagnole, Canal of the Bourne, Canal of the Rhone, Canal of Vesubie, Pierre-latte canal, Canal of Manosque, Canal of the Herault, other late works.

C.—ITALIAN IRRIGATION LEGISLATION AND ADMINISTRATION.

CHAPTER IX—ITALY (¹); *Right of Property in and Control over Water-courses and Water-sources.*

INTRODUCTION—*Importance of the study of irrigation experience in Italy:*—The valley of the Po, the classic land of irrigation; Magnitude, number, and excellence of its irrigation works; Long continued systemization of its irrigation practice; Its irrigation customs crystallized into well ordered codes of laws.

SECTION I—BASIS OF PROPERTY RIGHTS IN WATER-COURSES AND WATERS IN NORTHERN ITALY:— Barbaric rule of the middle ages; The birth and development of the Italian republics; The rise and fall of the feudal system; The principles of the Roman law handed down in the customs of the people; The earliest known laws.

Government ownership of all natural streams of importance as irrigation feeders; The rule alike in Piedmont and in Lombardy, and now for all Italy; Declarations of ownership, in the royal ordinance of 1817; Instructions to intendants, of 1828; Sardinian code of 1837; And the new Italian code of 1865; Neither navigability nor floatability the test of public importance of a stream in Italy; The volume of waters available for irrigation, the test; The underlying physical cause for the difference between this rule in Italy and that of France.

Government control of water-courses; General regulations for water-courses in Piedmont (1817); Articles of the Sardinian Penal code, applicable to the affairs of water-courses (1837); River regulations in Lombardy; Special, for the River Lambro (1756 and 1782).

SECTION II—OWNERSHIP AND CONTROL OF SPRINGS:—Character, number, and great importance of the springs in the valley of the Po; Resemblance of these *fontanili* to the *cienegas* of southern California; Private property right in springs, and acquired rights to use spring waters, in Lombardy; In Piedmont, the articles of the Sardinian code; Comparison of these with the analogous articles of the French code; In the present kingdom of Italy, the articles of the Code Victor Emmanuel; Comparison of these with those of the Sardinian code.

Regulation of the opening of springs; Origin and source of the spring waters; Necessity for restrictions on the opening of new springs; In Lombardy, the law of 1804 and the decree of 1806; Comments of De Buffon on the foregoing; In Piedmont, the articles of the Sardinian code; Regulations present questions for experting; Opinion of Giovanetti.

SECTION III—THE RIPARIAN RIGHT:—No private streams, except small rivulets, in Italy; No control of waters by riparian proprietors on public streams; Riparian rights on private streams, under the Sardinian code; Comparison of these articles with analogous provisions of the French code; The articles of the new Italian code on this subject; Comparison with those which preceded it.

CHAPTER X—ITALY (²); *Water Privileges and Canal Works, and the Administration of Waters and Works.*

SECTION I—THE RIGHT TO CONSTRUCT WORKS IN AND TO DIVERT WATERS FROM STREAMS:—Governmental policy in regard to water privileges; In ancient Milan, and Venice; In modern Lombardy, and Piedmont; In the present, unified Italy; Applications and formalities for water privileges; Piedmont, instructions of 1828; Lombardy, and the present kingdom of Italy; Terms of water-right concessions; Lombardy, regulations of 1806; Piedmont, articles of the Sardinian code; All Italy, articles of the Code Victor Emmanuel.

SECTION II—ADMINISTRATIVE REGULATION OF WATER-COURSES:—The general administrative organization of Italy; The local administrative organizations of the departments; River regulation in Piedmont; Instructions to the agents of the domain; Regulations of 1817; For navigable, and for non-navigable rivers; General river regulations for Lombardy, promulgated by the Austrian rulers, for the province of Mantua.

SECTION III—ADMINISTRATION OF GOVERNMENT CANALS:—Organization of the administrative bureau; In Piedmont under the minister of finance; The office of works, and the engineering corps; Instructions to the agents of the domain, concerning canals and waters; Piedmontese system; General regulation for the administration of the royal canals; The Lombardian system; The system for all Italy.

CHAPTER XI—ITALY (³); *Regulation of Irrigation Practice.*

SECTION I—DISTRIBUTION AND MEASUREMENT OF WATERS.—Hydraulic science and practice; The problems of distribution and measurement; The Piedmontese legislation, articles of the Sardinian code; Legislation for all Italy, articles of the Code Victor Emmanuel; Remarks on these provisions; First system—distribution by volume; Opinions of Giovanetti, De Buffon, and Sclopis; Importance of settled conditions; Second system—distribution by use or service; Third system—distribution by time.

SECTION II—THE RIGHTS OF IRRIGATORS:—(1) The right to a continuance of water supply from canals; A great struggle over this point; The former ruling in Piedmont, not recognized in the Sardinian code; The struggle between canal men and irrigators in Lombardy under Austrian rule; Long leases and carefully drawn agreements, the outcome of these contentions; (2) The right to the use of spare waters; Contentions over this point in Piedmont; The case of the Marquis de Saint G.; Articles of the Sardinian and of the Italian codes, on this point.

SECTION III—OBLIGATIONS AND RIGHTS OF IRRIGATORS AND CANAL MEN:—(1) Concerning water supply and use; Provisions of the Sardinian and of the Italian codes; (2) Priority of privilege in distribution; Schedules for distribution; The Sardinian and Italian rulings.

CHAPTER XII—ITALY (⁴); *Regulation of Drainage and Works connected with Irrigation Practice.*

SECTION I—REGULATION OF WORKS ACCESSORY TO IRRIGATION PRACTICE:—Distances to be preserved from boundaries of tracts; The necessity for regulations on this point; The articles of the Sardinian and of the Italian codes; Obligations concerning the construction and maintenance of works; Prevention of interference; Articles of the codes.

SECTION II—THE RIGHTS AND OBLIGATIONS OF DRAINAGE:—Necessity for drainage in Italy; Troubles arising out of drainage matters; Opinions of Baird Smith; The principles of the Piedmontese law; Articles of the Sardinian code; The law of Lombardy; Articles of the Italian code; Comparisons of these laws.

CIVIL LAW COUNTRIES: ITALY.   11

SECTION III—SANITARY LEGISLATION :—The unheeded teachings of experience ; Evil effects of unregulated and unskilled irrigation ; Legislative regulation of rice culture in Lombardy and in Piedmont; The question one for general legislation, but also for administrative judgment; Modern sanitary legislative regulations, in connection with irrigation of rice and of meadows; In Lombardy, a special decree of 1809; In Piedmont, a general law of 1855.

CHAPTER XIII—ITALY ([5]); *The Right of Way to conduct Waters.*

SECTION I—SOME ANCIENT AND MODERN LAWS ON THE RIGHT OF WAY SUBJECT :—The Milanese code of 1216; The Venetian code of 1455; The Charles Emmanuel code of 1770; The Lombardian laws and decrees of 1804 and 1806; The wisdom of these last laws overlooked by the Austrian rulers of Lombardy in 1816 ; Trouble growing out of this oversight; Decision of the Aulic council at Vienna and return to the Napoleonic laws in 1820.

SECTION II—THE SERVITUDE OF WAY TO CONDUCT WATERS :—Piedmont under the Sardinian code; Nature of the right; Form and amount of compensation; Three forms of the right-of-way question; First form—the right of aqueduct across lands; Articles of the Sardinian and Italian codes; Noteworthy points in these provisions; Compensation ; Second form—the right to cross other canals; Sardinian and Italian codes' provisions, and comments thereon; Third form—the right of aqueduct by a common channel ; Sardinian and Italian codes' provisions, with remarks thereon ; The right of aqueduct for waters of drainage and for warping ; Lombardian, Piedmontese, and general Italian laws.

SECTION III—RIGHT OF AQUEDUCT FOR PUBLIC WATERS :—Condemnation for purposes of public utility ; The Sardinian and Italian codes' provisions, with remarks thereon ; Favorable opinions of Smith and of DeBuffon on the system.

CHAPTER XIV—ITALY ([6]); *Irrigation Organization and Regulation.*

SECTION I—IRRIGATION ORGANIZATION :—Causes and necessities for organization; Social tendency of irrigation in Italy; Formation of irrigation associations in Lombardy; General law of association in Lombardy.

SECTION II—ORGANIZATION AND MANAGEMENT OF IRRIGATION ASSOCIATIONS :—The general association of irrigation west of the Sesia; Piedmont; Internal organization and management; The direction-general and the council of arbitration ; Finance and superintendence ; Relations of the government and the association; Government lease of waters and canals to the association; Rights and privileges under the lease; Management of the waters and maintenance of the works; Water-power and mills; Revenue and rents.

SECTION III—ORGANIZATION OF IRRIGATION ASSOCIATIONS :—The present law for all Italy; Voluntary associations of landholders; Compulsory formation of associations; The principles recognized as to inseparable community of interests and public utility in such works.

CHAPTER XV—ITALY ([7]); *Irrigation Enterprise.*

SECTION I—FORMS OF ENTERPRISE AND EXAMPLES OF CANAL CONSTRUCTION :—The association principle not generally applied in carrying out main canal works ; Review of principal works carried out in ancient and modern times; The great modern work, the Cavour canal ; Its character, size, location, cost, history, and unfortunate management.

SECTION II—CONCESSIONS TO CAPITALIZED COMPANIES :—Analysis of the concession to the Cavour canal company; (1) Obligations of the company; (2) Conditions of the concession ; (3) Privileges to the company ; (4) Benefits to the company.

SECTION III—GOVERNMENT POLICY AND ENCOURAGEMENT TOWARDS IRRIGATION :—General policy as to public works; Prize competition in irrigation practice; The royal decree of 1879; The hydrographic survey of Italy.

D.—Spanish Irrigation Legislation and Administration.

Chapter XVI—Spain (¹); *The Right of Property in and Control of Water, and Water-courses.*

INTRODUCTION.—IMPORTANCE OF THE STUDY OF THE IRRIGATION SYSTEMS OF SPAIN:—The physical conditions of, and the necessity for irrigation in Spain: The great central plateau of Spain; The south and east coasts; Rainfall and its distribution through the year; Resemblance of these conditions to those presented in California; Similarity in cultivations also; The necessarily fragmentary treatment of the subject.

SECTION I—ORIGIN OF PROPERTY RIGHTS IN SPAIN:—Barbarian rule in Spain; Dominion of the Goths; The codification of the Gothic laws; The inroad and dominance of the Moors; Conflicting local laws and customs; Expulsion of the Moors; Local customs and rights left firmly implanted; Gradual unification of Spain; Repeated codification or compilation of the Spanish laws; The principles of the Roman laws as to waters, molded upon a Gothic form, and modified in application by Moorish customs and local administrative organization; Recent codifications.

SECTION II—OWNERSHIP AND CONTROL OF WATERS AND WATER-COURSES:—Principles of the Spanish law; The Institutes of the Civil law of Spain; Classified division of property; Public property and common property; The communal system of the Roman provinces; Communal property rights; The administration of communal property; Ancient community water-rights and irrigation enterprise.

SECTION III—GOVERNMENTAL ADMINISTRATION AND REGULATION OF WATER-COURSES AND WATERS:—Governmental organization of Spain; The ministry of *Fomento*; The division of ports, canals, waters, etc.; The civil engineering bureau; The provincial governments; Communal administrative system of modernized Spain; The public-works policy of Spain; River improvement and guarding.

Chapter XVII—Spain (²); *The old general Water-laws of Spain.*

SECTION I—RIVERS AND RIVER WATERS, AND THE UTILIZATION OF THEM:—Rivers as distinguished from torrents; Public ownership of rivers; Construction of works in and on the banks of rivers; ownership of waters of rivers; Rights to divert waters; Navigable and non-navigable rivers, and non-navigable tributaries of navigable rivers; Extent and nature of riparian rights; Diversion and use of waters encouraged under government regulation; Instructions to governors of provinces, in 1788; Royal decree to promote irrigated agriculture, in 1819; Rights under these laws; Governmental regulation of these privileges; Royal order of 1839.

SECTION II—SMALL STREAMS AND TORRENTS, AND THE UTILIZATION OF THEIR WATERS:—The distinction between public and private waters; Local and communal control and utilization; Waters on private estates; Extent and nature of riparian rights; The rights of the owner of one bank of a private stream; The rights of the owner of both banks; Construction of works in private streams; Principles as to division of waters; Relative rights of large and of small estates; Relative rights of upper and lower riparian proprietors; Transfer of riparian privileges; Acquirement of rights by prescription; Division of riparian properties; Additions to riparian properties; Ownership of the beds of streams; The right to rain waters on private lands, and on public roads.

SECTION III—OWNERSHIP AND CONTROL OF SPRINGS:—The principles of the Roman law: Private property in springs on private estates; Public ownership of springs on public lands; Control of spring waters on private estates; Acquired rights to use spring waters; Acts which constitute prescription; Conflicting views on this point discussed; Opinions of Escriche, Lopez,

CHAPTER XVIII—SPAIN (³); *Old local Water-laws and Customs in Spain.*

SECTION I—OWNERSHIP AND CONTROL OF NATURAL STREAMS UNDER ANCIENT GRANTS:—The case of the river Turia in Valencia; The grant by King James I of Aragon, in 1238; Administration of the rights; Ownership and control in Murviedro, in Almansa, in Alicante, in Elche, in Murcie, in Lorca, Nijar, Grenada, and a number of other localities; Under old grants and customs.

SECTION II—WATER RIGHTS AND LANDHOLDINGS:—The water-right systems as connected with landholdings in each of the localities mentioned above.

SECTION III—THE OUTCOME OF THE SYSTEM OF LOCAL CONTROL OF STREAMS:—Conflicting water interests; Irrigation enterprise paralyzed; The safeguard of the union of land and water and the inalienability of water rights.

CHAPTER XIX—SPAIN (⁴); *Old local Irrigation Regulations and Customs.*

SECTION I—IRRIGATION ORGANIZATIONS:—The syndicate associations of Valencia, of Murviedro, of Almasa, of Alicante; of Elche; of Murcie; of Lorca, and other localities.

SECTION II—IRRIGATION REGULATIONS AND ADMINISTRATION:—The various rules and regulations and the administration thereof in the above named and other localities.

SECTION III—SPECIAL WATER TRIBUNALS:—Organization and administration of justice in irrigation affairs by the special water tribunals of Valencia, Lorca, and other localities.

CHAPTER XX—SPAIN (⁵); *The new general Law of Waters for all Spain. (Analysis of.)*

SECTION I—RIGHT OF PROPERTY IN AND RIGHTS TO THE USE OF WATERS:—A—Waters upon private property: (1) Rain waters and the waters of torrents; Rights to use and works for diversion; Limitation as to extent of utilization. (2) Waters of springs and other sources; Ownership of source waters, private; Loss of control of such waters; The old and the new law; Extent of right under the old law; Water power and irrigation rights; Priority of right in time of drought; Obligations to return the water to its natural course. (3) Waters of streams or rivulets on private property; Private streams defined; Utilizations of waters; Riparian rights; Priority of privileges; Estates crossed and estates touched by the stream; Nature of prior rights; Perfection of acquired rights; As between utilizers, first in time first in right; Rights of the source owner; Forfeiture of his rights; Riparian proprietors only can utilize waters of private streams; Return of surplus and unused waters; When public waters become private. (4) Subterranean waters; The waters of ordinary wells; Protective distances between wells; Artesian well waters; Acquirement of rights to use artesian waters; Rights to sink artesian wells on private lands. B—The waters of the public domain: (1) Rain waters and the waters of torrents Privileges to use the waters of public torrents, arroyos, road drains; Prescriptive rights to use these waters; Construction of works for the diversion of these waters; Liability of those who erect such works; Waters of intermittent springs; Rights for storage works. (2) Drainage waters from towns and public establishments, and waters discovered or developed on public works. (1) Public ownership of these waters; Acquirement of rights to use; Riparian privileges with respect to these waters. (3) The waters of public sources, springs, rivulets, and rivers; What waters naturally public; When waters become public; Acquirement of right to use public waters; When official sanction necessary, navigable and non-navigable rivers.

SECTION II—GRANTS OF RIGHTS TO THE USE OF PUBLIC WATERS :—(1) Official sanctions or concessions; Sanctions from the governors; Non-navigable rivers, when machinery is not operated by steam; Limitations as to amounts; Government concessions. (2) Proceedings to obtain official sanctions and grants or concessions; Formalities as to applications, presentation of cases, publications of intention, hearing of objections, examinations of projects, reports to central administration; Authority of governors; Decisions of the administration; Subsidies to irrigation enterprises; Auction sales of concessions; Grants without subsidies; Pledges, securities, inspections, limitations as to time. (3) The terms of special grants or concessions; Protection of prior rights; Priority of privileges; Development of subterranean waters; Responsibility of new grantees as to damages to old rights; Forfeiture of grant by non-compliance; Surplus drainage waters, public. (4) Grants for the use of public waters in irrigation :—Order of preference in the use of public waters; Provision as to change of use; Duration of grants; Prevention of waste of water; Continuous flow; Prior rights; Lands for headworks. (5) Rights to develop subterranean waters on public lands:—Municipal sanctions; Government sanctions; Ownership of waters; Applications and formalities; Authority of governors; Security deposits; Reservations with grants; Final concessions; Ownership of waters; Forfeiture of rights.

SECTION III—THE RIGHT OF WAY FOR WATERS:—The right for public canals; The right for private canals; Cases wherein the right will be enforced; Applications to be made; The right by a common channel not allowed; Objections and opposition to rights of way proceedings; Permanent and temporary rights; Compensation; The right across roads, other canals, etc.; Character of channels; Occupation of lands during construction; Works at crossings; The enlargement of constructed canals; Protection of rights; Repairs and maintenance of works; Ownership and control of canals; Rights of land owners; Determination of width of old canals; Forfeiture of rights of way; Annulment of rights of way; Disposal of forfeited works; Rights of way in cities; Rights of way by private agreement; Rights to construct a dam on private lands; Construction of works in canals.

SECTION IV—IRRIGATION, ORGANIZATION, AND ADMINISTRATION:—The province of the administration; Government regulations; Authority of the administration; Jurisdiction of the courts.

Irrigating communities:—When communities may be formed; Syndicates to be elected; Laws and rules to be adopted; Memberships and qualifications; Representation of interests; Duty and authority of syndicates; General assemblies; Financial assessments; Admission of new members; General syndicates.

Tribunals of Irrigation:—Composition of tribunals; Functions and powers of tribunals; Proceedings of tribunals; Ancient tribunals to be undisturbed.

Special privileges to irrigation enterprises:—Power of obtaining materials; Occupation of public lands; Exemption from security deposits; Exemption from taxation of works; Special privileges; The land tax remains the same for ten years; Subsidies; Water leases and obligations of irrigators.

CHAPTER XXI—SPAIN ([6]); *Governmental Policy, and Irrigation Enterprise.*

ANALYSES OF SEVERAL RECENT GENERAL LAWS CONCERNING CANAL CONCESSIONS AND THE UTILIZATION OF PUBLIC WATERS IN IRRIGATION.

Outline of subject-matter and formulation of the law of February 20, 1870.

Formalities to be observed in applications for and the issuance of permits to utilize public waters in irrigation.

Duties of communal, provincial, and general administrative authorities in the matter of the issuance of permits for the utilization of waters.

CIVIL LAW COUNTRIES: SPAIN. 15

Ordering of consideration of applications; relative priority of applicants and claimants.

Guarantees of good faith and ability to perform and carry out the works required, and conditions of forfeiture of grants preliminary to carrying out works.

Commencement, carrying forward, and completion, or forfeiture (in case of non-completion), and sale (in case of forfeiture) of works.

Benefits accorded to concessionary companies; Tax subsidy; Collection of tax subsidies, etc.

Benefits, etc.:—Indemnities from interest; Exemptions from taxation, etc.; Declarations of public utility.

Benefits to private enterprises:—Exemptions from taxation on increased values, etc.

Examinations of irrigation projects, and reports on the same. How paid for.

Reservations as to companies already formed and protection of existing rights by concession.

[NOTE.—The foregoing law repeals and amends a portion of the general law of 1866 of which an abstract is given in the preceding chapter.]

A Royal decree of December 20, 1870, giving instructions at great length and in detail, for the administration and observance of the forms and rules laid down in the preceding law of February, 1870.

[NOTE.—Abstract of contents omitted here because of great length.]

An Order of the Governor of the Republic, concerning the establishment of Juries of Irrigation, or special tribunals to consider and decide irrigation questions: dated March 20, 1873.

An Order of the Governor of the Republic, concerning the establishment of a department within the Ministry of Public Works for examination and reporting on hydraulic questions of public interest: dated March 29, 1873.

An order of the Governor of the Republic concerning the execution and observance of the General Law of waters of August, 1866: dated April 5, 1873.

A General Law concerning the useful employment of public waters in Irrigation and otherwise, amendatory of the law of August, 1866: dated June 13, 1879.

A Royal decree concerning a project for a new law governing concessions to water companies: dated November 17, 1879.

A project for a new law for direct subsidies to companies carrying out irrigation works: dated November 17, 1879.

A Royal order concerning the application of several articles of the law of June, 1879: dated September 5, 1881.

A Circular of the Director General of public works to the Governors of provinces concerning the employment of public waters in irrigation: dated September 7, 1881.

[NOTE.—Later legislation of importance sent for and expected.]

CHAPTER XXII.—SPAIN ([7]); *Governmental Construction and Management of Irrigation Works.*

Abstracts of the laws, decrees, and orders relative to the establishment, construction, maintenance, government, and operation, of the great public works of irrigation in Spain; the Royal canal of *Aragon*, the Imperial canal of *Tauste*, and others.

[NOTE.—Further reference to the contents of this chapter is omitted because of its considerable length and the impossibility of outlining it adequately in a small space.]

CHAPTER XXIII.—SPAIN ([8]); *Construction and Management of great Works of Irrigation by Companies and Societies.*

Abstracts of special laws and concessions of right to construct irrigation works by companies and societies; giving for Spain the same class of information as is given for France in Section II of Chapter VIII, preceding.

E.—Mexican Irrigation Legislation and Administration.

Chapter XXIV—Mexico (¹); *Ownership and Control of Waters and Water-courses.*

INTRODUCTION—General character of Mexico as an irrigation country; Characteristics of Mexican irrigation enterprises and customs; Former connection of the country with, and the implanting of customs and rights in California.

SECTION I—FOUNDATION OF PROPERTY RIGHTS, AND RIGHTS TO WATERS AND WATER-COURSES:—The conquest of Mexico and early Spanish policy in Mexico; Forms of land titles and nature of land holdings; Early-day agriculture in Mexico.

SECTION II—SPANISH LAWS RELATIVE TO MEXICO:—Laws concerning the colonization of New Spain, and the settlement of the Indies; The *"Recopilacion de leyes de las Reynos de las Indias";* The laws governing settlements; The provisions concerning irrigation and waters for all purposes; Instructions to *corregidores;* Special instructions to leaders of expeditions to form settlements, pueblos, missions, etc.

SECTION III—THE MEXICAN REPUBLIC AND THE SPANISH LAWS:—The independence of Mexico; The retention of the Spanish laws relating to waters and water-courses and to lands; Some general statutory laws of Mexico concerning the subject; Local growth of customs relating to irrigation and the control of streams; Some State statutory laws relating to irrigation and the regulation of water-courses.

Chapter XV—Mexico (²); *The General Law of Waters for all Mexico.*

SECTION I—OWNERSHIP AND CONTROL OF WATERS AND WATER-COURSES.
SECTION II—WATER RIGHTS AND PRIVILEGES; THEIR ACQUIREMENT AND FORFEITURE.
SECTION III—MEASUREMENT AND DISTRIBUTION OF WATERS AND THE CONTROL OF IRRIGATION.

[NOTE.—Under the above general headings this chapter contains a rearrangement of the general law of waters for Mexico, as given by GALVAN in his *Ordinanzas de Tierras y Aguas.*]

Chapter XXVI—Mexico (³); *State and local Irrigation Customs and Regulations.*

SECTION I—COMMUNITY ORGANIZATION AND PUEBLO RIGHTS AND PROPERTIES.
SECTION II—STATE AND LOCAL WATER REGULATIONS AND ADMINISTRATION THEREOF.
SECTION III—IRRIGATION CUSTOMS.

[NOTE.—Under the above headings this chapter contains the information for Mexico, so far as available, analagous to that already given for Spain, under similar title lines, in preceding chapters.]

BOOK II.

IRRIGATION LEGISLATION AND ADMINISTRATION IN COUNTRIES UNDER THE COMMON LAW.

THE LAW OF WATERS AND WATER-COURSES, SO FAR AS IT AFFECTS IRRIGATION QUESTIONS, TOGETHER WITH THE STATUTORY LAWS, THE CUSTOMS, AND GOVERNMENTAL POLICIES WITH RESPECT TO IRRIGATION, IN COUNTRIES WHERE THE COMMON LAW PREVAILS AND IN THE ENGLISH COLONIES.

INTRODUCTION.—The origin and development of the Common Law, and the physical, political, and social circumstances which have molded it as respects the subjects of this report.

CHAPTER I—ENGLAND ([1]); *The Right of Property in Water-courses and Waters.*

CHAPTER II—ENGLAND ([2]); *Governmental Control, Improvement, and Regulation of Water-courses.*

CHAPTER III—ENGLAND ([3]); *Natural or Riparian Rights to Water and the Control of Water-courses.*

CHAPTER IV—ENGLAND ([4]); *Acquired Rights and Privileges to the use of Waters from public and private sources.*

CHAPTER V—ENGLAND ([5]); *Governmental Policy with respect to Water supply, and the use of Waters for all purposes.*

CHAPTER VI—ENGLAND ([6]); *Irrigation in England; Its character, extent, purpose, and relation to Water supply and Water-courses.*

CHAPTER VII—INDIA ([1]); *Ownership and Control of Water-courses and Waters.*

CHAPTER VIII—INDIA ([2]); *Governmental Policy and Private or Corporate Irrigation Enterprise.*

CHAPTER IX—INDIA ([3]); *Governmental Enterprise; Construction and Management of Irrigation Works.*

CHAPTER X—INDIA ([4]); *Irrigation Regulations and Customs.*

CONTENTS: BOOK II; IRRIGATION LEGISLATION.

CHAPTER XI—NEW ZEALAND; *Ownership and Control of Water-courses; Governmental Policy in Waters Conservancy and Irrigation.*

CHAPTER XII—VICTORIA; *Ownership and Control of Water-courses; Governmental Policy in Waters Conservancy and Irrigation.*

CHAPTER XIII—NEW SOUTH WALES; *Ownership and Control of Water-courses; Governmental Policy in Waters Conservancy and Irrigation.*

CHAPTER XIV—OTHER COLONIES; *Ownership and Control of Water-courses; Governmental Policy in Waters Conservancy and Irrigation.*

CHAPTER XV—UNITED STATES OF AMERICA (1); *Ownership and Control of Water-courses and Waters.*

CHAPTER XVI—UNITED STATES (2); *The use of Waters on the Public Domain.*

CHAPTER XVII—STATES AND TERRITORIES (1); *The Legislation of Texas, New Mexico, Arizona, Nevada, etc.*

CHAPTER XVIII—STATES AND TERRITORIES (2); *The Legislation of Colorado, Montana, Idaho, Wyoming, etc.*

CHAPTER XIX—CALIFORNIA (1); *Ownership and Control of Water-courses and Waters; The Mexican law and the Common law in California.*

CHAPTER XX—CALIFORNIA (2); *The Legislation of Water-rights and Water-courses.*

CHAPTER XXI—CALIFORNIA (3); *The Legislation of Irrigation and Kindred Subjects.*

CHAPTER XXII—CALIFORNIA (4); *Irrigation Customs, Regulations, and Enterprise in California.*

CONCLUSION AS TO LEGISLATION.

CHAPTER XXIII—*The Water-right Conflict in California.*

CHAPTER XXIV—*Subjects for Legislation with respect to Irrigation in California.*

CHAPTER XXV—*A Water-right System for California.*

CHAPTER XXVI—*An Irrigation System for California.*

[NOTE TO BOOK II.—Under the above given headings, this book treats the subject of Irrigation legislation, custom, and administration for the countries mentioned, in the same manner, with the same degree of fullness of detail, and in the same spirit, as Book I treats the subject for countries where the Civil Law prevails.]

BOOK III.

WATER SUPPLY AND IRRIGABLE LANDS IN CALIFORNIA.

A.—PHYSICAL FEATURES OF CALIFORNIA.

CHAPTER I—*The Mountains, Hills, and Valleys of the State.*

GENERAL OROGRAPHICAL FEATURES:—The Coast Range of mountains from the northern boundary of the State to the San Francisco Bay region; the region west of the Coast Range and north of the Harbor of San Francisco: the Monte Diablo Range and the Coast Range, south of the Harbor of San Francisco to the Sierra de San Rafael in Santa Barbara and Ventura Counties; the Bay region; the region west of the Monte Diablo Range; the Sierra San Rafael, Sierra Madre, San Bernardino, and San Jacinto range of mountains, from San Luis Obispo County to the southern boundary of the State; the region west and south of the last named range of mountains.

The Sierra Nevada range of mountains; the great Interior or Central Valley of California, composed of the Kern, Tulare, San Joaquin, and Sacramento Valleys; the region north of the Sacramento Valley, and east of the Coast Range; the region east of the Sierra Nevada Mountains, from the Oregon border to the Lake Tahoe basin; the Tahoe basin; the region east of the Sierra Nevada Mountains, and south of the Tahoe basin to the Mojave desert; the Mojave desert; the Colorado desert.

NOTE.—Under the above headings this chapter will contain, within about 15 to 20 pages of printed matter, a general description of the geography and topography of the State, written expressly with the view of laying the foundation for an understanding of the more detailed descriptions, and the discussions of the subjects of water-shed areas, rainfall, drainage, water-courses, water supply for irrigation, and irrigable lands, which are to follow. This chapter refers for illustration to the Topographical map of California—scale 30 miles in the inch—one of the set of small scale general maps of the State which have been prepared for the atlas collection, and as accompaniments of this report. It should be illustrated by a double page wood-cut map of the State, reduced from that above named, and bound in with the matter of this volume.

CHAPTER II—*The Rainfall, Drainage, and River Systems of California.*

GENERAL HYDROGRAPHICAL LAWS OF THE PACIFIC COAST REGION; Influences of mountain ranges on rainfall in California; Collection of statistics of rainfall; Construction of a rainfall chart for the State; Distribution of rainfall in the State; Maximum and minimum periods of rainfall; Years of ordinary precipitation; Years of flood and years of drought; Necessity of much more extended and widespread observation; A State weather service, similar to those of Iowa, Ohio, and other States, working with the United States Signal Service bureau.

The water-shed areas of California, and the streams which drain them; Of the region draining into the ocean north of the San Francisco bay region; Of the coast region draining into the San Francisco and San Pablo bays; Of the region draining into the ocean south from the bay of San Francisco to Point Conception; Of the region draining into the ocean south of Point Conception; Of the region draining into and of the Central Valley of the State; Of the region draining east from the Sierra Nevada mountains; Of the region draining into the Mojave desert; Of the region draining into the Colorado desert; The Colorado river.

NOTE.—Under the foregoing headings this chapter will contain, within 25 to 30 pages of printed matter, a general discussion of the subject of rainfall in the Pacific Coast region, and particularly in California; an account of rainfall statistics collected in and for California, and of the making of a rainfall chart from them; together with a presentation of the summarized results of this study. It will also contain a general description of the individual watershed zones of the state—their altitudes, areas, general characters, exposures, rainfall, and drainage; a general description of the main streams which collect from these areas—their characters, courses, and volumes of flow (approximately); and a general review of the subject of water supply for irrigation in the state at large.

The description of the drainage areas of the state, while, of course, not based upon complete data, is written from much information never before compiled, and, for the general purposes of the report, will be quite satisfactory at least for the regions where the study has been carried forward by this department, as indicated in the chapter following.

The rainfall statistics referred to in this chapter, and upon which much of its matter is based, are by far the most complete and extended tables ever collected for the state, and will be annexed, in a summarized form, to this book as an appendix.

The present chapter will refer for illustration to the General Drainage-Area Map of the state—scale, 12 miles in the inch—and to the Rainfall Chart—scale, 30 miles in the inch—the former prepared specially as an illustration of this subject, and the latter for a like purpose and also as one of the set of atlas sheets for independent publication. The letter press should be illustrated by several double and single page wood cuts prepared from these maps and from numerical data on hand.

In the final writing of this chapter it may be found advisable to divide its matter and make two chapters of it.

CHAPTER III—*The Water supply for Irrigation.*

THE REGION WHEREIN OBSERVATIONS HAVE BEEN MADE BY THE STATE ENGINEERING DEPARTMENT:—The nature, extent, and value of those observations: Methods of observation and of compilation of results; The general subject of observation and recordation of the flow of streams.

THE GREAT CENTRAL VALLEY OF CALIFORNIA:—Its streams, and their floods and failures of supply; The streams which enter the Sacramento valley from the Coast Range south of Red Bluffs; The Sacramento river; The streams which enter the valley from the Sierra Nevada mountains north of Feather river; The Feather river; the creeks between Feather and Yuba rivers; The Yuba river; The creeks between Yuba and Bear rivers; The Bear river; The creeks between Bear and American rivers: The American river; The creeks between American and Cosumnes rivers; The Cosumnes river; The streams between the Cosumnes and the Mokelumne rivers; The Mokelumne river; The streams between the Mokelumne and Stanislaus rivers; The Stanislaus river; The creeks between the Stanislaus and Tuolumne rivers; The Tuolumne river; The creeks between the Tuolumne and Merced rivers; The Merced river; The streams between the Merced and the San Joaquin river; The San Joaquin river; The creeks between the San Joaquin and Kings river; The Kings river; The creeks between the Kings river and the Kaweah; The Kaweah river; The streams between the Kaweah and Kern river; The Kern river; The creeks which enter the San Joaquin valley from the Coast Range.

THE LOS ANGELES AND SAN BERNARDINO VALLEYS:—Their streams and water supply: The Los Angeles river; The San Gabriel river; The Santa Ana river; Cañons intermediate between these rivers.

NOTE.—Under the foregoing specific headings, within the space of 25 to 30 printed pages, this chapter will contain a summarization of the information concerning the flow of streams—flood, low, and ordinary—which has been collected by observations, cursory and definite, made by the State Engineering department during the period of its existence, and collected from reliable sources.

This chapter will refer to an appendix for tabulated data in detail, and, for illustration to the General Drainage-area Map.

It should be illustrated by a number of wood cut diagrams, graphically presenting the data arranged in tabular form in the appendix.

It is, perhaps, needless to say that although one of the least interesting to the general reader, this is one of the most important chapters of the work.

CHAPTER IV—*Underground Water supply for Irrigation.*

THE SUBTERRANEAN FLOW OF STREAMS; Identification of waters; Development of waters; *Cienegas*, or flowing marsh-springs of Southern California; Artesian wells and water supply.

NOTE.—Under the above specific headings, for the several regions of the Sacramento, San Joaquin, Tulare, Kern, Santa Clara, Los Angeles, and San Bernardino valleys, and some other limited localities, will be grouped, within about 25 to 30 pages of printed matter, the substance of the data collected on the subject of underground water supply and its development; together with discoveries of the problems involved, and the deduction of some conclusions which may be of special value.

The chapter will refer to an appendix for detailed descriptions of the results of work in the several quarters; and should be illustrated by a number of wood engravings prepared from the numerical and other data given, and also for illustration of the discussions advanced.

B.—WATERS, LANDS, AND IRRIGATION IN CALIFORNIA.

CHAPTER V—*The Hydrographic Districts of the State.*

THE REGIONS WHERE THE WATER SUPPLY IS, IN EACH, MEASURABLY INDEPENDENT OF THAT IN OTHERS:—The region of the Sacramento valley; The region of the San Joaquin valley; The region of the Tulare and Kern valley; The region of the San Bernardino and Los Angeles valleys.

NOTE.—Under the above general headings, within the compass of 25 to 30 printed pages, will be given in this chapter a general description of the irrigable portions of the regions named, of the stream channels through them, and the topographical relation of these water supply sources and the lands to which the waters may be conducted. This account is preliminary to a more detailed discussion of the subject of irrigation districts contained in the next chapter.

The present chapter will refer to the contour line topographical map of the great central valley of the State—scale, six miles in the inch; to the special river maps of the valley, and to the topographical map of the San Bernardino and Los Angeles valleys, for illustration; as also to numerical data in tabular form found in an appendix.

CHAPTER VI—*Irrigation Districts.*

THE REGIONS WHICH MAY BE IRRIGATED TO THE BEST ADVANTAGE AND ECONOMY BY ONE SET OF WORKS AND UNDER ONE MANAGEMENT:—The character of the irrigable lands; The probable cultivations of the future; The duty of water; The probable irrigation works of the future; The extent of possible irrigation; Irrigation of mountain valleys; Irrigation of foothill regions; Irrigation of reclaimed swamp lands.

Probable clashings of interest; Probable unification of interests; Gradual development of irrigation districts; Probable future changes in their outlines; Impracticability of a rigid system.

NOTE.—The foregoing general headings indicate with sufficient clearness the character of matter which will make up this chapter. It will be a general discussion of the subject of outlining irrigation districts, as they are required or desired to be formed in the state; will be based on the data, as to lands, water supply, duty of water, cultivations, etc., collected by the department; and will relate specially to the Sacramento, San Joaquin, Tulare, Kern, Los Angeles, and San Bernardino valleys.

The matter will occupy 25 to 30 pages of print; and will refer to the topographical maps of the valleys named, and of the rivers in those valleys for illustration.

CHAPTER VII—*Water-rights in the Irrigated Districts of California.*

EXISTING CLAIMS TO WATER, AS RECORDED AND PUBLICLY ANNOUNCED; EXISTING APPROPRIATIONS OR DIVERSIONS OF WATER; EXISTING USES OF WATER.

NOTE.—Under the above general headings, for the regions of the San Joaquin, Tulare, Kern, Los Angeles, and San Bernardino valleys, will be given the general results of the work of examining the records of water claims, the measurements of the water diversions, and the observations of extent of use of waters, made by the State Engineering department.

The chapter will take about 25 to 30 pages of printed matter, and will refer to several extensive appendices in which will be found the data from which it has been written.

C.—The Future of Irrigation in California.

Chapter VIII—*Physical Effects of Irrigation*(¹).

IRRIGATION AND THE ARTERIAL DRAINAGE OF THE COUNTRY:—Irrigation and the prevention of floods; Irrigation and navigation; Irrigation and reclamation; Irrigation and the debris problem; The reproduction of waters.

NOTE.—Under the above general headings, there will be found in this chapter a discussion of those physical effects of irrigation which closely concern the water supply problems and the usefulness or efficiency of the natural streams of a country; together, in all cases, with a direct application to the practical questions which will come up in California as irrigation is extended. The local applications are made upon the basis of the surveys of our streams, the study of the water supply, and of the question of demand likely to be made for irrigation purposes.

Chapter IX—*Physical Effects of Irrigation*(²).

IRRIGATION AND LAND DRAINAGE:—The rising of soil waters; Irrigation and soils; Physical effects on soils; Chemical effects on soils; Changes in character of plant growth; Irrigation with muddy or silt charged waters; Systematic *colmatage*, or warping of lands; Irrigation and underground water supply.

NOTE.—Under the above general headings, this chapter will contain a discussion of those effects of irrigation which pertain to the lands irrigated, and the necessities which thereby arise; as, also, some mention and description of the working of other uses of water in ways analogous to irrigation methods. The practical application will be made to the several chief irrigation regions of our State, and illustrations will be introduced from home experiences.

Chapter X—*Physical Effects of Irrigation* (³).

THE CLIMATIC AND SANITARY EFFECTS OF IRRIGATION:—Humidity and temperature of the air; Rainfall, and irrigation; Sanitary effects of irrigation; Origin of malarial influence; Influence of soils, subsoils, and cultivations; Unhealthful effects of certain cultivations; Influence of waters; Irrigation with muddy waters; Irrigation with sewage waters.

NOTE.—Under the above general subject headings, the present chapter will contain a discussion of those effects of irrigation pertaining to healthfulness of the irrigated districts, together with their practical bearing on the future of irrigation in the several regions for irrigation in California.

Chapter XI—*Conservation of Waters for Irrigation, in California.*

PRESERVATION AND EXTENSION OF FORESTS; The influence of forests on water supply; The destruction of forests in California; Reforesting of mountains; The necessities of the future; The system of the future; Storage; Economy of delivery.

NOTE.—In this chapter will be found a discussion of the subject of forestry and water supply based upon the data collated by the best authorities, and the opinions of the leading writers on the topic. The legislation of other countries will be summarized and a system sketched out for application under our conditions and political system.

Chapter XII—*The Practical Problems of Irrigation reviewed for California.*

FUTURE EXTENSION OF IRRIGATION:—Irrigation populations; Irrigation and capital; Water right systems; Sale and distribution of waters; Irrigation enterprise; Necessary systemization of irrigation and arterial drainage.

NOTE.—Under the above subject headings the whole problem of irrigation as presented for the future in California will be reviewed, in the light of the study of its legal, social, political, financial, and physical elements, which have now been gone through with.

BOOK IV.

EXISTING, PROJECTED, AND PROBABLE WORKS AND IRRIGATIONS IN CALIFORNIA.

CHAPTER I—*Early Irrigation in San Bernardino and Los Angeles counties.*

OLD SPANISH AND OTHER SETTLEMENTS:—The Pueblo of Los Angeles; The Mission of San Gabriel; The Mission of San Fernando; The Mormon settlement of San Bernardino; The Mexican settlements on the San Gabriel; Early utilization of the waters of *cienegas*; Other small local works.

NOTE.—The valleys of Los Angeles and San Bernardino were the scenes of irrigation practice long before the acquirement of this territory by the United States, and as these were the pioneer irrigations, worthy of note, in the state, I devote a chapter to a brief sketch of their history.

CHAPTER II—*Irrigation in San Bernardino valley.*

The works deriving waters from the Santiago, Temescal, Ucuipa, and San Timoteo creeks.
The works deriving waters from Mill creek and the Upper Santa Ana river.
The works deriving waters from Plunge, City, Twin, Devils cañon, Cajon, and Lytle creeks.
The works deriving waters from Sainsevain's, Day's, Smith's, Reid's, and Clark's cañons, and Cucamongo, and San Antonio creeks.
The utilization of waters from *cienegas* and artesian wells in San Bernardino valley.
The cultivations and irrigations in the settlements of Riverside, Redlands, Old San Bernardino, San Bernardino, Lytle creek, Etawanda, Ontario, Cucamongo, Pomona, and others.
New or projected works and irrigations in and for the San Bernardino valley district.

CHAPTER III—*Irrigation in the Upper San Gabriel valley.*

The works deriving waters from Puente creek, and San Dimas and Dalton cañons.
The works deriving waters from the Upper San Gabriel river.
The works deriving waters from the Fish, Sawpit, Santa Anita, Little Santa Anita, Bailey, Davis, Precipico, and Arroyo Seco cañons.
The utilization of waters from *cienegas* and artesian wells in the Upper San Gabriel valley.
The cultivations and irrigations in the settlements or neighborhoods of Puente, Asuza, Santa Anita, El Monte, San Gabriel, Pasadena, and Sierra Madre.
New or projected works in and for the Upper San Gabriel valley district.

CHAPTER IV—*Irrigation in the San Fernando and Los Angeles valleys.*

The works deriving waters from the Arroyo Seco, Verdugo, and Tejunga cañons.
The works deriving waters from the Los Angeles river.
The utilization of waters from *cienegas* and from artesian wells in the Los Angeles valley.
The irrigations and cultivations in the pueblo of Los Angeles.

The irrigations and cultivations in the several other neighborhoods in the valleys of Los Angeles and San Fernando.
New or projected works in and for the valleys of San Fernando and Los Angeles.

CHAPTER V—*Irrigation in the Lower San Gabriel river district.*

The works deriving their waters from the Old river.
The works deriving their waters from the New river.
The utilization of waters from *cienegas* and artesian wells, Lower San Gabriel district.
The cultivations and irrigations in the several settlements of Lower San Gabriel district.
New or projected works for the Lower San Gabriel river district.

CHAPTER VI—*Irrigation in the Lower Santa Ana river district.*

Irrigation works from and northwest of the Santa Ana river.
Irrigation works from and southeast of the Santa Ana river.
The utilization of waters from *cienegas* and artesian wells in Lower Santa Ana district.
The cultivations and irrigations in the several settlements of the Lower Santa Ana district.
New or projected works for the Lower Santa Ana river district.

CHAPTER VII—*Irrigation in several other localities of the extreme Southern counties.*

Irrigation works, practice, projects, in several districts of San Diego, San Bernardino, Los Angeles, and Ventura counties, not mentioned in the preceding six chapters.

CHAPTER VIII—*Irrigation in Kern county.*

Irrigation works and practice of and from Kern river.
The works in the district between "Old South Fork" and "Old River" channels of Kern river.
The works in the district between Old River channel and the present channel of Kern river.
The canal works in the district lying between Kern river channel and Goose lake slough.
The canal works in the district known as Swamp land Districts No. 121, 184, 185, and 208.
The canal works in the district north of Kern river and the district third above named.
The canal works in the district south of Kern river and the district first above named.
The artesian wells in Kern county and the utilization of waters from them.
The cultivations and irrigation practice in the districts of Kern county above named.
Projected works and possible extension of works and irrigation in and for the districts named.

CHAPTER IX—*Irrigation in Tulare county, exclusive of that from Kings river and that in the Mussel Slough district.*

The canal works and irrigations in the district south of the Kaweah river irrigations.
The canal works and irrigations deriving water supply from the Kaweah river, but exclusive of those for the Mussel Slough country.
The artesian wells and irrigations therefrom in the districts above described in Tulare county.
The cultivations and irrigation practice in the districts above described.
Projected works and possible extension of works and irrigations in the districts above described.

CHAPTER X—*Irrigation in Tulare county, from Kings river and in the Mussel Slough district.*

The "Seventy-six" and other Upper Kings river south side works.
The People's, Mussel Slough, Last Chance, Lower Kings River, and Rhoads' ditches in the Mussel Slough country.
The Settlers and Lakeside ditches in the Mussel Slough country.
The artesian wells in the Mussel Slough country.
The cultivations and irrigation practice in the districts supplied by the works above named.
Projected works and possible extension of works in the districts above named.

CHAPTER XI—*Irrigation in Fresno county from the Kings river and San Joaquin river, and in the district between them.*

The Kings River and Fresno canal and the irrigation therefrom.
The Fresno canal company's canal and the irrigation therefrom.
The Centerville and Kingsburg company's canal and the irrigation therefrom.
The Fowlers Switch company's canal and the irrigation therefrom.
The Emigrant company's canal and the irrigation therefrom.
The Liberty canal and the irrigation therefrom.
The Upper San Joaquin River company's canal works and projected extensions thereof.
The cultivations and utilizations of water, generally, in the districts of the above works.
Projected works and probable extension of old works and irrigation in the districts commanded by the above named works.

CHAPTER XII—*Irrigation on the east side of the San Joaquin valley, between the San Joaquin and Merced rivers.*

The Chowchilla canal and the irrigations therefrom with waters of the San Joaquin river.
The canal works and irrigations deriving supply from the Fresno river.
The canal works and irrigations deriving supply from the Bear, Mariposa, and Dry creeks.
The Farmers' canal, deriving water from the Merced river, and the irrigations therefrom.
The artesian wells in the region between the San Joaquin and the Merced rivers.
The cultivations by irrigation in the region above described.
Projected and possible works and probable extension of works and irrigations in this region.

CHAPTER XIII—*Irrigation on the west side of the San Joaquin valley.*

The San Joaquin and Kings River canal company's works.
The irrigations from the San Joaquin and Kings river canal.
The cultivations by the waters of the above canal.
Artesian wells, and the utilization of the water thereof.
Projected or possible works for the extension of irrigation in this region.

CHAPTER XIV—*Irrigation in Stanislaus and San Joaquin counties.*

The canal works and irrigations from the Merced river on the north side.
The canal works and irrigations from the Tuolumne river.
The canal works and irrigations from the San Joaquin river.
The canal works and irrigations from the Calaveras river.
The canal works and irrigations from the Mokelumne river.
Works and irrigations with waters of the Cosumnes river.
Artesian wells and utilization of the waters thereof in the region above described.

The cultivations by irrigation in the region above described.
Projected or possible new works and probable extension of existing works.

CHAPTER XV—*Irrigation in the Sacramento valley (east side).*

Works and irrigations with waters of the American river.
The Natoma Land and Water company's works and irrigations.
The North Fork of American company's canal, and the irrigations therefrom.
Works and irrigations with the waters of Yuba river.
The Excelsior canal company's works and irrigations.
The works and irrigations from the creeks in Butte and Tehama counties.
The Stanford canal near Vina and the irrigations therefrom.
Cultivation by irrigation in the regions above described.
Projected and possible works, extensions of works, and the possibilities of irrigation in the region above described.

CHAPTER XVI—*Irrigation in the Sacramento valley (west side).*

Works and irrigations deriving waters from Cache creek.
Projected works from Cache creek and possible extension of Cache creek irrigation.
Projected works of irrigation from Putah creek and possible irrigation from this source.
Projected irrigation from Stony creek and possible irrigation from this source.
Projected or possible works from the Sacramento river.
Utilization of well waters in irrigation within this region.
Cultivation by irrigation within this region.

CHAPTER XVII—*Irrigation from mining ditches in the foothill and mountain regions of Mariposa, Tuolumne, Calaveras, Amador, El Dorado, Placer, Nevada, Sierra, Plumas, and Butte counties.*

[NOTE.—The data for this chapter is not complete; but will be quite full for El Dorado, Placer, and Nevada counties, and parts of Butte, Amador, and Calaveras.]

CHAPTER XVIII—*Irrigation in Shasta, Siskiyou, and Modoc counties.*

[NOTE.—There has not been any data collected for this chapter.]

CHAPTER XIX—*Irrigation in Mono and Inyo counties.*

[NOTE.—There has not been any data collected for this chapter.]

CHAPTER XX—*Irrigation in Santa Barbara, San Luis Obispo, Monterey, Santa Cruz, and San Benito counties.*

[NOTE.—There has not been any data collected for this chapter.]

CHAPTER XXI—*Irrigation in Santa Clara and Alameda counties.*

[NOTE.—There has not been any data collected for this chapter.]

CHAPTER XXII—*Summarized Statement as to existing Irrigation Works and possible Extension of Irrigations in California.*

[NOTE.—This chapter will be full so far as the preceding chapters furnish the information.]

BOOK V.

THE DISTRIBUTION OF WATERS AND THE PRACTICE OF IRRIGATION.

CHAPTER I—*Irrigation in its relation to Agriculture.*

The province of irrigation, with respect to its immediate fertilizing and moistening effect on soils and plants, and other results sought by the agriculturist, considered in its relation to different soils, cultivations, climates, and peoples, and especially with respect to Californian circumstances.

CHAPTER II—*Methods or Systems of Irrigation.*

The ways of preparing lands for and conducting irrigation work, considered with respect to quantity of water supply, extent of lands, character of soil, nature of cultivation, peculiarities of climates and agricultural populations, and general customs of people.

CHAPTER III—*Cultivation of special Crops by Irrigation.*

The necessity for, utility of, preparation for, and the conducting of irrigation, with respect to special crops or cultivations—such as fruits, vineyards, vegetables, cereals, meadow grasses, other forage plants, sugar cane, rice, etc.—and under varied conditions of soil, climate, extent of water supply, and agricultural populations.

CHAPTER IV—*The Duty of Water in Irrigation.*

A compilation and summarization of the known results of irrigation, as to the question of the area of lands irrigated by any given unit measure of water, or, conversely, the quantity of water required to irrigate any given area of land during a recognized irrigation season, under the varied circumstances of soil composition, drainage, crop cultivation, climate, preparation of lands, method of irrigation, and delivery of water; together with a discussion of the principles involved in the problems of the economical use of water.

CHAPTER V—*The Cost and actual Value of Irrigation (gravity systems).*

A compilation and summarization of the known results of irrigation as to the cost of works per acre irrigated; the cost of water per crop, per area served, or per unit of measure, when bought; the cost of preparing lands for irrigation; and the cost of conducting the operations of irrigation per season, and per acre, or per crop; and also a compilation of statistics as to the financial returns from irrigation enterprise, works, and practice, in the matters of crop production and increased land values.

CHAPTER VI—*The Cost of Raising Water for Irrigation.*

A summarization of information available as to the cost of pumping water for irrigation, by steam, wind, horse, and water power, with tables for data applicable for estimating on such cost under varying conditions of delivery, and cost of fuel, etc.

CHAPTER VII—*Methods of Distribution and Delivery of waters for Irrigation.*

A discussion of the various systems of selling, leasing, dividing out, and delivering waters—as by volume, by service, and by time or season—together with a compilation of rules of delivery, of administration, and of management of works in partitioning out waters to customers, in vogue in California and many other irrigation countries.

CHAPTER VIII—*Measurement of flowing Waters.*

A description of the various methods of measuring or estimating the amount of water flowing in open canals and aqueducts, over weirs and falls or drops, through sluices under pressure, and in pipes, with illustrations of apparatus required for simple cases, and with tables to facilitate calculations.

CHAPTER IX—*Modules and Partitioners for dividing and delivering Waters.*

Descriptions and illustrations of the various devices designed and used in different countries in meting out waters for irrigation, where it is desired to deliver a constantly uniform flow; and also in cases where it is desired to divide a flowing stream into proportionate parts at all stages of its flow, together with a discussion of the principles upon which these are designed, and the value to be placed upon them to accomplish their purpose in practice.

CHAPTER X—*General Results of Irrigation Practice.*

A description of results usually attending long-continued irrigation practice, in the matters of effects on drainage, rising of soil waters, reproduction of waters in springs or channels at lower levels, rising of ground levels from deposits of silt by the waters used, leaching out of soils, rising of alkali from subsoils, etc., together with citation of noted examples of each, and rules for overcoming to some extent the bad effects produced.

BOOK VI.

WORKS FOR THE INTERCEPTION AND STORAGE OF WATER FOR IRRIGATION.

CHAPTER I—*Sites and Dams for Reservoirs.*

A classification of reservoir sites according to the topographical features of a country. Short period interruptions—as in the *presas* of Mexico, the *estanques* of Spain, and the *bunds*, or shallow tanks of India. Long period storage in deep reservoirs. Mountain valley and cañon, hill-land valleys, and other sites considered. The various kinds of dams suitable for the several sites and purposes spoken of. Essentials of an efficient dam, and the importance of a large margin of safety in planning dams. Materials for the construction of dams; their examination and selection. Sites and foundations for dams; their examination and selection.

Chapter II—*Earthwork Dams and Embankments.*

Simple embankments of earth of various kinds; Earthwork dams with puddle walls, and with spiling walls. Selection of sites for earthwork dams. Examination and preparation of foundations. Examination and selection of materials.

Planning and proportioning of embankments. Processes of construction. Finishing off of embankments; Paving, graveling, turfing, planking, etc.

Examples of great and small embankments and dams, with illustrations of plans and statistics of cost, so far as available.

Chapter III—*Timber Dams.*

Dams in which timber is the chief component material: Framed or cribbed timber bolted down, and with planking; The same also loaded with rock, gravel, or other ballast. Selection of sites for such dams. Preparation of foundations, etc.

Planning, proportioning, processes of construction, and finishing off of such works.

Examples of dams of these descriptions, with illustrations and statistics.

Chapter IV—*Rock Dams, of the Californian Mining type.*

Dams in which roughly quarried rock masses, laid without timber cribwork, and without coursing or cementing, forms the body of the structure. Selection of sites and materials. Examination and preparation of foundations.

Planning, proportioning, processes of construction, and finishing of dams of this class.

Examples and illustrations of dams of this class, statistics of cost, so far as available, etc.

Chapter V—*Masonry Dams; Spanish, French, and Belgian.*

Dams composed of stonework laid up solidly, well bedded and cemented.

Sites, foundations, materials, and preparations thereof for dams of this class.

Planning, proportioning, laying out, construction, finishing, etc., of such works.

Examples and illustrations of the great dams of this class in the world, with statistics and accounts thereof, so far as available.

Chapter VI—*Outlets, Gates, and Galleries to Reservoirs.*

Methods and plans, with illustrations, for drawing the waters off from reservoirs, through, under, or around dams of the several kinds described, in or on foundations of the various characters found in practice.

Chapter VII—*Failures of Reservoir Dams.*

A discussion of the causes of failure of reservoir dams, with illustrations, and accounts of a number of cases of great failures.

Chapter VIII—*Cost of Reservoir Works.*

A compilation and summarization of the cost of storing water in reservoirs so far as statistics are available.

The comparison of the costs of dams of the various kinds described, with reference to a number of storage sites examined in California.

Sources of and amount or proportion of loss of water stored. Evaporation and percolation. Efficiency of a storage work with allowances for shortage in delivery.

Chapter IX—*Accounts, Plans, and Statistics of great Storage Reservoirs.*

The great "tanks" of India and Ceylon; The "pantanos" of Spain; The "presas" of Mexico; The reservoirs of England, California, and elsewhere in the United States.

BOOK VII.

WORKS FOR THE DIVERSION, CONDUCTING, AND DISTRIBUTION OF WATERS IN IRRIGATION.

CHAPTER I—*Dams for the Diversion of Water from streams.*

The several classes of dams for this purpose; Their adaptability under various stated circumstances; Their use in different countries; Their relative advantages and disadvantages.

The general practice as to works of this kind in India, Italy, Spain, France, and some examples in California.

CHAPTER II—*Immovable, Overfall Dams.*

Dams immovable, and built of various materials—such as brush with rock or gravel; timber and lumber with rock, gravel, or sand; rock or stone without cement or mortar; rock, stone, or brick with cement or mortar.

Foundations for and adaptability of each of these works. Planning and construction.

Examples and illustrations of noteworthy works of the several kinds, now in existence.

CHAPTER III—*Movable Dams.*

Dams made so as to be automatically or otherwise removable for the purpose of leaving a free waterway during floods.

Double, falling, shutter dams of the Indian type.

The Poireé and the Chanoine iron-frame and wooden-needle dams.

The Desfontaine drum weir.

The Thenard and Mesnager double shutter weir.

The Kraus pontoon weir, etc.

Adaptability of these works; Their cost, planning, and construction.

CHAPTER IV—*Headworks and Regulators for Canals.*

The various kinds of works applicable and used for governing and regulating the flow of water into canals.

General planning and disposition of such works at the heads of canals and in connection, or not, with dams of diversion.

Descriptions and plannings for works of this kind built of various materials, sizes, and forms.

Descriptions and illustrations of notable works of this character in India, Italy, Spain, and elsewhere.

CHAPTER V—*Canals and Ditches.*

The exploration for, planning, laying out, and construction of canals and ditches.

The questions of capacity, sectional dimensions, grade slopes, velocity of flow, transportation of sediment, erosion of beds, silting, etc., considered.

IRRIGATION WORKS IN AND FOR CALIFORNIA.

Classification of canals according to use and purpose, with discussion of the characteristics of each class.

Descriptions and illustrations of a number of notable existing works, great and small, in other countries, and of some in California, also.

CHAPTER VI—*Aqueducts and Flumes.*

Open channels, other than cut and embanked canals, for conducting and conveying water.

Aqueducts of stone, brick, cement, and iron; their designing, laying out, and construction.

Flumes of wood; their adaptability, designing, and construction.

Notable examples of works of these several kinds now existing in France, Italy, India, or in California.

Descriptions, illustrations, statistics of construction, and cost.

CHAPTER VII—*Pipes and Syphons.*

Closed channels or conduits for conducting and conveying water under pressure.

Syphons of brick, stone, or cement masonry for conveying the waters of one canal under the channel of another water-course.

Iron pipes for conveying water under pressure across depressions.

Pipes of iron, wood, earthenware, cement, etc., for the conveying and distribution of water in irrigation.

Descriptions and illustrations of a number of notable constructions of these classes.

CHAPTER VIII—*Regulators, Drops, and Overfalls.*

Constructions designed for and placed in the channels of canals to regulate their flow by diminishing the grade slopes of their waters above and admitting of sudden falls in the waters at the structures themselves, so as to start the channels at lower planes below.

Classification and description with respect to materials and class of construction and volume of water to be handled.

Description and illustrations of notable works designed and serving for this purpose in the great canals of India, Italy, and Spain.

CHAPTER IX—*Outlets and Modules.*

Works for the drawing off of water from canals and distributaries, and for meting out stated volumes of flow in distribution.

The classes of these works with respect to the service to be performed and the materials to be employed in their construction.

Theory and practice of design and construction.

Descriptions and illustrations of structures in use in various irrigation regions.

CHAPTER X—*Operation and Maintenance of Works.*

The operations of caring for, cleaning, repairing, and otherwise maintaining in good condition, canals, aqueducts, and other irrigation works.

Machinery and appliances used in these operations.

Regulations and customs governing these operations in the great irrigation regions.

CHAPTER XI—*Artesian Wells.*

The construction of artesian wells, as practiced in France and other European countries, and in California.

Descriptions and illustrations of the apparatus and machinery used in construction. Statistics of artesian well construction and cost, so far as available.

EXTRACT FROM INTRODUCTORY LETTER.

To his Excellency GEORGE STONEMAN, *Governor of California:*

SIR: *   *   *   *   *   *   *

In the preparation of this report, it is sought to keep constantly in view the legitimate purpose of the work; namely, the promotion of agricultural prosperity by irrigation in California.

It is believed that this is to be accomplished by the establishment of a thorough general understanding of The Problems of Irrigation on the part of those who have to do with its practice, and those who are charged with the making of laws to foster and control its development.

The subject is a great one, presenting many phases. It has its legal, social, political, economic, physical, technical, and practical problems. He who would understand this subject must look well to these, one and all.

There is a vast fund of experience had in other countries, that carries its general lessons, which we cannot, with reason, neglect; but these must be studied systematically, else we be led into errors by overlooking some governing conditions not apparent to the less thorough observer.

The literature holding these data is very voluminous, for the most part in foreign languages, and itself far from systematiized. It is simply a great labor to collect and go through with it and cull out, compare, judge of, and arrange its useful materials, and draw and apply practical lessons from these.

In addition to the study of the irrigation questions as founded on apparent conditions around us here, I have endeavored to bring to our enlightenment, by the results of such labor, the legitimate outcomes of irrigation experiences elsewhere.

The ground which I have traversed has now been marked out in the foregoing table of contents. Some of the results are embodied in final form in the following advance sheets.

No attempt has been made to write a text-book for lawyers, a manual for engineers, or a complete guide to practice for irrigators, nor a treatise for the scientist or the political economist. But each will find in this report, when finished, very much which probably would not otherwise come to his notice, and so linked with the phase of the question of which he may make a special study, that he will be profited by the reading and prepared to be less uncompromising in his views.

Now, it is the eradication of uncompromising and unreasonable views of this irrigation question which is necessary to the attainment of its solution. No mere local or class study of it can effect this purpose, and no one person, unless specially devoted to it, can go over it all with the material scattered and undigested as it heretofore has been.

In undertaking this report the writer has believed that its cost could only be returned to the people of the State by making it sufficiently thorough to constitute a guide to the whole subject. A report only on what might be called the practical or engineering problems of irrigation in California, while probably of much use to a very few persons, would have been of little use to the State or the people at large.

This question will be a living one, growing in importance, and pressing for legislative action in some form, for years to come; and this report has been framed and carried forward with a view of facilitating this action.

      *   *   *   *   *   *   *

Very respectfully, your obedient servant,

WM. HAM. HALL, State Engineer.

SACRAMENTO, CAL., December 31, 1884.

REPORT OF THE STATE ENGINEER.

PART I.

IRRIGATION LEGISLATION AND ADMINISTRATION.

BOOK I.—COUNTRIES UNDER THE CIVIL LAW.

The Roman Empire; France; Italy; Spain; Mexico.

CHAPTER I.—THE ROMAN EMPIRE;

ROMAN LAWS AND ADMINISTRATIVE POLICY WITH RESPECT TO WATERS AND WATER-COURSES.

INTRODUCTION.—Time and Circumstances of Forming, and Importance of Roman Laws of Waters.

SECTION I.—Right of Property in Waters and Water-Courses.
Common Property—Running Waters.
Public Property—Rivers; Private Property—Brooks.
Resumé as to Ownership.

SECTION II.—Control of Public Rivers and Waters.
Construction and Maintenance of Works.
Use of Public Rivers.
Use of Public Waters.

SECTION III.—Control of Waters in Private Works, and on Private Lands.
Private Springs.
Ownership of Waters of Springs and in Works.
Acquirement of Right to use Waters.

SECTION IV.—Right of Way to Conduct Waters.
Prædial Servitudes—*Aquæ ductus; Aquæ haustus.*
The Right on Private Property.
The Right on Public Property.

INTRODUCTION.

It may be said that Rome once ruled all the countries of Southern Europe, Northern Africa, and Western Asia where irrigation had its birth and its greatest development in ancient times;* and that her laws with respect to waters were crystallized several centuries after the Romans became familiar with the practice of irrigation and the necessities of the irrigation interest in the various quarters of this region.§

---

* Irrigation of course existed in some of these countries long before the Roman Empire was founded, and India and China also were the scenes of irrigation practice at a much earlier period.
§ Rome was all powerful throughout the Mediterranean countries before the Christian era; but the Theodosian codes were promulgated more than four centuries, and the Justinian codes more than five centuries later.

Those who were regarded as authority at the law in Rome, who plucked from the confusion of her earlier customs and edicts the principles of her laws, with others who expounded those principles and formulated her system, were amongst the most acute and logical thinkers the world has known to this day; so that modern jurisprudence, at least in continental Europe, has been so far guided by the principles of the Roman Law, it has been said, in substance, that "having ceased to rule the world by their arms, the Romans still control mankind by their reason."

This being the case, it is well that our inquiry commence with a glance at the leading features of the laws and administrative policy of this people in their dealings with the water-right and irrigation interest, although, considering the vast difference in our social establishment and forms of business enterprise, we may not find the positive guide which we are in search of.

SECTION 1.

RIGHT OF PROPERTY IN WATER AND WATER-COURSES.

In their classification of things, as a basis of laws regarding ownership and use, the Roman jurists recognized, with respect to propertyship, two general classes:

Things *in patrimonio*, capable of being possessed by persons exclusively of others; and things *extra patrimonium*, those incapable of being so possessed.

Things *extra patrimonium* were classed under four headings:

Things common, free to all mankind; things public, belonging to some nation or people; things *universitatis*, belonging to some certain city, society, or corporation; and things *nullius*, belonging to nobody; the latter relating to things consecrated and devoted to religious uses.

COMMON PROPERTY—RUNNING WATERS.

Like the air, water was regarded as a necessity to human life, of which every one might use so much as was requisite for personal requirements, but which was not capable of appropriation to private ownership further than in this sufficient quantity.

By the law of nature, flowing water is a common property of all men.—[Justinian's Codes, Lib. 6, Tit. 1, Sec. 1.

"*Res communes*, * * * things the property of no one in particular * * *—the air, running water, the sea and its coasts, and wild animals in a state of freedom. The air is necessary to human life, and every one may use so much of it as is requisite, but it is not capable of appropriation; the same is the case with running water."—[Colquhoun, § 923.

"*Res omnium communes.* Such things, it is obvious by their very nature, could not stand in private ownership. Every person might use and enjoy them, but no one could possess them. These things are the air, running water, etc. * * * * * *

"When the Romans speak of the air as a *res omnium communis*, they do not mean to include the space above the earth, but only the atmosphere. The man who owns the soil owns the space above it, and this space is a thing *in commercio*"—(capable of barter or sale); but the atmosphere is a *res extra commercium*"—(a thing not capable of barter or sale).

"The same remarks apply to running water. The space in which the brook or streamlet flows, as it hastens to feed the larger streams, is in private ownership, but the water is not."—[Gaius, p. 209.

"Things common to all, are those which being given by providence for general use cannot be reduced to the nature of property. Such are the air, running water, the sea, and the shores of the sea; but if a man by prescription, from time immemorial, had the use of running water, as for a mill, his case was an exception to the general rule, but he must not waste the water unnecessarily; and mills and other structures might be erected on rivers by special license. Vid. Digests, 48–8."—[Browne, Vol. I, p. 170.

"From the very nature of such things results the necessary consequence that they can never be completely the object of private ownership, that they can form the object of such a right only so far, and so long, as it is possible for man to retain them under his dominion or control. Except as to the portions which an individual may thus have brought under subjection, they must be regarded as common to all the world—*Res omnium communes.*"—[Goudsmit, p. 113.

PUBLIC PROPERTY—RIVERS: PRIVATE PROPERTY—BROOKS.

Streams, rivers, lakes, ponds, etc., which were not in private ownership, were regarded as public things, and spoken of as *res publicæ*, things which belonged to the people as a nation.

There were public properties used for State purposes, solely and only, by the representatives of the State, the rulers or officials; and public properties used by private individuals, and yielding revenue to the State for such use; and there were public properties used freely by all the people.

"*Res publicæ*, in the strict sense of the words, are those things which are exclusively in the possession of the State. Such are public thoroughfares, public streams, public squares, public baths, and the amphitheatres."—[Gaius, p. 210.

The roads and rivers were specially counted as public things by the Romans. "The public could use the river, for instance, as a ship way, or for fishing, but the ownership itself was vested in the State."—[Gaius, p. 210.

They were not the property of the ruling sovereign, but of the sovereign power of the people collectively, each one of whom could use them as his own, but might not injure them, neither segregate any portion or constituent part of them for his own.

And this right of use in the navigable rivers, highways, harbors, and gates was extended to all, whether Roman citizens or not, who were at peace with Rome.

Public rivers are defined to be such as were perennial or ever flowing, as distinguished from winter torrents, but this, although one of the essentials of such rivers, was not alone sufficient to render them public, for if located through private lands they were not the property of the public unless navigable or capable of being made so by improvement, or, from some other cause, of public importance.

"It is not, however, all streams that are public things. Thus Ulpinius says: 'Some streams are public and some are not. Cassius defines a public river as one which runs perennially.' A perennial stream is one which flows throughout the year.

"Perennial brooks are not as such *res publicæ*, although, in consequence of their resemblance to public streams, legal protection was afforded to persons having only a private interest in them, which protection was based upon and analagous to that by which waters that were *res publicæ* (public) were protected.

"There was not at any time in Roman law a strictly legal distinction drawn between the river (*flumen*) and the rill or brook (*rivus*). As a general thing, it may be said that the brook is a private thing, and the river the property of the public."—[Gaius, p. 209.

A river (flumen) was distinguished from a stream (rivus) by its greater volume, or more considerable local importance. Rivers were of permanent flow, or only of intermittent flow, leaving their beds dry in Summer, when they were called torrents. A permanent river might occasionally dry up, however, without losing its character. Permanent rivers were public rivers, and might be either navigable or not navigable.*

RIVER BANKS AND BEDS—OWNERSHIP AND USE.

The bank of a river, like the shore of the sea, commenced at the limit of the spread of the waters at high stage, but when lands were not inundated; land above that line was property in public or private ownership; all below that line was the bed of the river.

In the case of navigable rivers and all streams of the public property, the beds belonged to the State; being part of the public thing—the river. Should the waters leave such channel and take another, the river, the public thing, was considered to have moved, and the old bed became the property of those whose lands were taken for the

---

*Justinian D., Lib. 43, Tit. 12, Sec. 1; Lib. 43, Tit. 12, Sec. 3.

new channel, while lands taken for this new channel became part of the public property—the river.

In the case of non-navigable rivers and streams not regarded as public, situated on private property, the beds belonged to the riparian proprietors. While these beds were covered with water it was considered that the rights of such proprietors were suspended, but such rights revived when the waters receded.

It is not necessary for the purposes of this report to carry this subject further and consider the matter of alluvion.

By some authorities, and at a different period of time, a somewhat different doctrine was held to, regarding the beds of rivers, which was as follows:

The beds of rivers were classed with animals, birds, and bees in a wild state, fishes in public water, gems unfound, etc., things capable of private ownership, but yet within the power or possession of no one.

When abandoned by the waters the lands of such beds became the property of the riparian proprietors, as did also alluvial formations in the beds of a stream—whether in the form of addition to the banks, or islands in the channel—as soon as deposited.

"Temporary inundation suspends, and continued inundation destroys the right of the owner."—[Colquhoun, § 982.

The banks of a public river might belong to the riparian proprietor, to the extent that he had the right to take the fruits, cut the bushes, and fell the trees which grew thereon, but not so as to prejudice the use of the river or its banks by the public.

The public had a right to the use of the banks of navigable rivers, so that a qualified ownership of the soil of such banks was all that could be acquired by private persons.*

The owner of lands which were bounded by a ditch or wall following near the bank, or by a public road on the bank of a public stream, was not a riparian proprietor; to be such his lands had to be bounded by the stream itself.

RESUMÉ AS TO OWNERSHIP.

We thus see, and it is essential to keep clearly in view, that the Roman law made a marked distinction between rivers and streams, and the waters thereof.

Taken as a whole, a river—bed and water—was regarded as a public thing (*res publica*), the property of the state, necessarily excluded from private ownership or control, barter or sale, the use of which in its entirety, to be enjoyed by all.

---

*Colquhoun classes the banks of navigable rivers amongst things public, and says expressly that they were public property so far as the public chose to use them in aid of navigation.

The water of the river was the property of all the people in common—it was regarded as susceptible of apportionment amongst the people—each might drink of it, each dip up a portion and carry it away, and, further than that, if the enjoyment of the public property—the river as a whole—would not be impaired, each might divert a portion of the water from its natural channel for other purposes than those of his own domestic necessities.

But the state, representing the people—the owner of the public thing, the river—was guardian of the common property, the water—and no person could use more than sufficient for his individual necessities and those of his family and cattle, without a special permit so to do.

Water sources and water-courses were susceptible of private ownership, and, where thus owned, the right to use their waters for purposes other than the supply of the immediate animal necessities pertained primarily to their possessor.

Thus, there were springs and brooks, which, being situated on private lands, constituted parts of such property, but the water itself, while running in its natural channel, was the property of all the people, and, as such, was the ward of the nation.

SECTION II.

CONTROL OF PUBLIC RIVERS AND WATERS.

It was specially declared to be lawful for every one to navigate his craft on all public rivers, lakes, and canals, and the banks thereof were open to all for purposes of loading and unloading, but the navigator was forbidden to enter forcibly upon a bank for this purpose.

The right of fishery was open to all, and each person might dry his nets upon the shore, and otherwise use the banks as might be necessary in the prosecution of his calling.

The banks and channels of public rivers were specially guarded from injury; the construction of works or the placing of obstructions therein, by the effect of which the current might be made more or less rapid, was forbidden.

The construction of works upon the bank, or in the channel of a public river, whether navigable or not, whereby either the low water or high water flow thereof would be affected, was forbidden.

And works which might have an effect such as described, erected without authority, were removed or abolished at the expense of the constructor.

"Prohibitory interdicts forbade anything being done tending to impede the navigation of public rivers, or changing the course of running water; and other interdicts, of the restitutory class, compelled the re-establishment of things in the way the public had hitherto enjoyed them."—[Ortolan-Mears, p. 398.

"The prætor says: 'I forbid any one to put any structure upon a river or on its banks, or to do anything that would deteriorate the navigation or the water-way.'"—[Justinian D., Lib. 43, Tit. 12, Sec. 1.

Speaking of this interdict, Colquhoun, in substance, says:

And this applied to all public streams, whether navigable or not, in full force except in the case of works intended for the protection or preservation of the banks or channel, the right to construct which works was the subject of a sanction.

"This interdict is intended for all people, and is perpetual, but 'lies against him only who has diverted the water, and his heirs prohibitorily and for restitution.'"—[Colquhoun, § 2291.

CONSTRUCTION AND MAINTENANCE OF WORKS.

It was declared to be lawful, however, for riparian proprietors, or those who lived near the bank of a public river, to erect works for the protection of a bank thereof, provided that navigation was in no way impeded thereby, and that the river or the other bank was not injured.

"The prætor does not pretend to prevent all kinds of works made in rivers or their banks, but only those which could injure navigation or the water-way. Thus the interdict of which we speak here only concerns the public rivers, and not the others."—[Justinian D., Lib. 43, Tit. 12, Sec. 1, § 12.

If damage resulted from any such work, an official examination was made, and, if deemed necessary, the works were removed, or ordered changed, and security for ten years was exacted from their owner or constructor, the amount thereof to be assessed by persons chosen for their competency in such matters.

There was an interdict, *de ripa munienda*, concerning the protection of river banks, whereby it was lawful for riparian proprietors to construct works for the repair or protection of the bank adjacent to their property. If damage was threatened by such works to the lands of another on the opposite bank or elsewhere, a writ of inquiry was ordered, and security was exacted for ten years against the results of such possible damage, if, in the opinion of experts, it was likely to occur.

"This interdict being only prohibitory, and not also restitutory, had to be applied for before the work was commenced; for, afterwards, there was no mode of making it effective, and recourse had then to be had, in case of damage done, to an action for damages."—[Colquhoun, § 2292.

DIVERSION OF PUBLIC WATERS.

Appropriation of the waters of public streams, except for individual use, was a custom not known to the Roman law, for although irrigation was recognized as a necessity, the rivers were regarded as a public property and as such were guarded in the common interest.

Navigable rivers and running waters generally were excluded from private ownership because of the public use to which they were devoted and the common necessity for their use.

The diversion of waters, whether of floods or low-water flow, from public rivers, reservoirs, or tanks, without the sanction of a special privilege in each case, was prohibited.

A decree of the prætor was required to obtain authority to appropriate to private use any material part of a property common to all the people.—[Ortolan, p. 143.

"Nothing prevents water being taken from a public river, unless the prince or the senate forbids; provided that this water may not be for public use. If the river is either navigable or makes another navigable, this will not be permitted."—[Justinian D., Lib. 43, Tit. 12, Sec. 2.

"The prætor must not accord the right of drawing from a navigable river a quantity of water whose extraction would injure navigation. It would be the same on a river which, not being itself navigable, discharged into another which it rendered navigable."—[Justinian D., Lib. 39, Tit. 3, Sec. 10, § 2.

"The matter of water, throughout the larger portion of the Roman empire, was a matter of great importance, and it was therefore found necessary to supply a summary remedy by interdict to all questions relating to it; hence it was provided in the edict : ' concerning annual water, it is not to be taken by force, fraud, or by the permission of another;' and 'concerning the use of Summer water, it is not to be taken by force, fraud, or by the permission of another.'"—[Colquhoun, § 2301.

"By the civil law, the rivers were public; * * * nor was any obstruction or diversion of a river allowed. See Digest, Lib. 43."—[Browne, Vol. 1, p. 171.

[See, also, extracts from Ortolan and from Colquhoun, given under the second heading preceding this one.]

It appears that water privileges were of two kinds : *First*—Those, to individuals, of water for use on individual lands—the terms "on his farm" being used in this connection; and these were accorded by local authority, apparently that of the provincial prætors, at one period at least. *Second*—Those of waters for public use, which authorizations emanated from the senate or other supreme central power.

When a joint right to divert was issued to several persons the matter of division of the waters was left to those holding the right.

The remodelling or alteration of the headworks of canals or cuts out from both public or private rivers, without official sanction, was prohibited.

USE OF PUBLIC WATERS.

The use to which water was to be put was not always stipulated in grants, provided that it was to be used in good faith and not wasted.

It was declared that the user of water was liable for damages, "by reason of anything done, dug, sown, delved, or built whereby the river was corrupted."

It was declared that water privileges should be "exercised in such a manner as not to damage other persons having similar rights."

"A *caput aquæ* was a head or source of water, where it first begins to appear in whatsoever manner."

All interference with public springs or water sources, lakes, wells, and fish ponds was prohibited.

"The waters of a public spring must be divided amongst the owners of the adjacent lands, in proportion to their possessions, unless some owner can prove his right to preference. But no one should be permitted to conduct the water on to his property unless it can be done without injury to others."—[Justinian D., Lib. 8, Tit. 3, Sec. 17.

The cleansing of springs or fountains, etc., was permitted, but it was stipulated that no new veins of water were to be opened up.

Reservoirs might be cleaned and repaired, but no additional waters conducted into them without authority.

Possession and use of running water, as for the operation of a mill, or in irrigation, by a private individual, from time immemorial, gave a prescriptive right to the continued enjoyment of such use.

No possessor of water, though having held it from immemorial time, had the right to use it wastefully to the prejudice of others.

SECTION III.

CONTROL OF WATERS IN PRIVATE WORKS.

Springs on private lands were the property of the land-owner, on the principles that to such proprietor belonged all above and all below the land, and all it produced.

The right to use spring waters might be acquired by others than the owner, by agreement or prescription; prescription being use, virtually, from time immemorial.

Spring waters flowing off, joining with other waters and forming brooks on other lands, became common property of all people, but their use was dedicated to the owners of the lands along their course; so that such waters, for purposes of diversion, belonged to these riparian proprietors.

It is necessary to carefully guard against misconception on this point. Water rising out of the ground on a private estate, as being a part of the spring, was the property of the owner of the land: he could do with it as he chose; but when any portion of that water had escaped from the tract where it came to the surface, it became a common property of all the people. But so long as it remained in channels on private estates and channels not public from any cause (navigability or other reason), only the owners of the banks of its channel could divert it from its course and use it, except this right should have been acquired as a servitude, as will be explained under the next heading.

But even these bank proprietors could not divert such waters, if, in doing so, other proprietors were injured thereby.

"For the validity of the concession for the right of taking water onto his property, it is necessary to have the consent not only of those in whose lands the water rises, but, further, of those who have the use of this water, that is to say, of those who have a right of servitude upon this water. * * * And, in general, it is necessary to have the consent of all those who have a right upon the stream or upon the land where the water rises."—[Justinian D., Lib. 39, Tit. 3, Sec. 8.

Water drawn from its source, diverted, or drawn from its course, into an artificial and private channel, or when stored in a reservoir or tank itself in private ownership, became private property.

The user might do with it as he chose, provided his use was in good faith—that he did not waste it.

SECTION IV.

THE RIGHT OF WAY TO CONDUCT WATER.

The rights to draw waters from a private spring or stream by others than its owner, and to conduct waters across lands owned by others, ranked as *servitudes*.

A *prædial servitude* under Roman law was a definite right of enjoyment in some particular respect, of one person's property by the owner of other adjoining or neighboring property. The land subject to the right was called *prædium serviens*, and the land to which the right was attached was called *prædium dominans*.

Such a servitude could be held only as an appurtenance to land owned, being called *prædial* because it could not exist without an estate.

The land subject to the servitude, and that to which the right of enjoyment was attached, had to adjoin each other, or be near to each other.

The servitude was attached to the land having the right of its enjoyment, and was owned with it, and passed to a new owner with the title to it; but was extinguished when the two estates involved, became the property of one person: that person then acting by right of absolute ownership of all the property, and not as owner of one estate and the attached servitude on the other.

Prædial servitudes related to estates in country or city, and were called, accordingly, *rural* and *urban*.

The right of passage across the lands of another, and the right of conducting water through such lands, appear to have been recognized as indispensable privileges from the earliest times of the Roman jurisprudence.

The right of way to construct a canal or other conduit through the property of another, and to lead waters through it, was known as *servitus aquæ ductus*, and was one of the chief rural servitudes.

"*Servitas aquæ ductus*, the right to convey water by canals, bricked trenches, or pipes through another's land. Some aqueducts were public, but others were for the use of private farms, to which latter this servitude particularly applies."—[Colquhoun, § 938.

The right to take water through the property of another in a ditch or other conduit, could be acquired by prescription—use for a long period of years—or by agreement, or, in the case of public works or works of public importance, title to the land necessary could be acquired by expropriation and payment therefor.

When acquired as a title, of course the right was complete. When, as a servitude, the right was acquired or accorded for a certain purpose only. Thus, he who had a prescriptive right to take any accustomed quantity of water across another's land, could not materially increase that quantity. Having taken the water for his own use, he could not take water also in the same channel for the use of another. Having taken water for a certain farm, he could not take more than enough for that farm.

"The quantity of water that could be taken was determined, in the absence of agreement, by custom, not by the wants of the land for which the servitude was granted; but so much could not be taken as to starve the land from which it came. If custom sanctioned it, the water might be used for irrigation."—[Hunter, following Justinian's Code, p. 245.

"No one can, without permission of the prince, conduct water across public property."—[Justinian D., Lib. 39, Tit. 3, Sec. 18, § 1.

A right to draw and use water from another's spring or rivulet might be imposed by agreement or prescription as a servitude thereon. This right was known as *aquæ haustus* and implied also the right of passage so far as necessary to exercise the servitude.

AUTHORITIES FOR CHAPTER I.

In the preparation of this paper I have consulted and compared the following authorities:

Colquhoun.—"A Summary of the Roman Law." By Patrick Mac C. de Colquhoun; 4 vols.; London—1849. See, more particularly, §§ 923, 924, 925, 938, 980, 981, 982, 983, 2289, 2290, 2291, 2292, 2300, 2301, 2302.

Goudsmit.—"The Pandects, a Treatise on the Roman Law." By J. E. Goudsmit, LL.D., 1 vol.; London—1873. See, more particularly, pp. 113-115 and foot notes.

Ortolan.—"The Institutes of Justinian." M. Ortolan. Analysis by T. Lambert Mears, M.A., LL.D.; 1 vol., London—1876. See, more particularly, pp. 143, 143, 220, 389.

Mackenzie.—"Studies in the Roman Law." By Lord Mackenzie; 1 vol., Edinburgh—1876. See, pp. 167-170, 177, 184-185.

Browne.—"Compendious View of the Civil Law." By Arthur Browne, LL.D.: University of Dublin; 2 vols.; London—1802. See, vol I, pp. 167-173.

Cumin.—"A Manual of Civil Law." Patrick Cumin, M.A., Baliol College, Oxford; 1 vol.; London—1865. See, particularly, pp. 80-85, 97-99.

Gaius.—"The Commentaries of Gaius on the Roman Law." Translated, etc., by F. Tomkins, M.A., D.C.L., and Wm. G. Lemon, LL.B.; 1 vol.; London—1869. See, particularly, pp. 201-210.

Justinian.—"Digest or Pandects,"—Of Justinian, Emperor. Translated into French by M. M. Hulot and others; 6 vols.; Metz—1804. See, particularly, books 8, 39, 40, titles quoted.

Hunter.—"The Roman Law in the Order of a Code." By Wm. A. Hunter, M.A., University College, London; 1 vol.; London—1876. See, particularly, pp. 168, 169, 241-245.

IRRIGATION LEGISLATION AND ADMINISTRATION.

FRANCE.

CHAPTER II.—FRANCE[1];

RIGHT OF PROPERTY IN AND CONTROL OF WATER-COURSES.

SECTION I.—*Origin of Property Rights and Ownership of Streams.*
    Basis of Property Rights in France.
    Ownership and Control of Navigable Streams.
    Ownership and Control of Non-navigable Streams.

SECTION II.—*Water Laws and Regulations.*
    Moving Causes of Development.
    Special Regard for Irrigation.
    Classification of Water Laws.

SECTION III.—*The Administration.*
    Administrative Purpose and Policy.
    Governmental Organization.
    The Administrative System.
    The Bureau of Public Works.
    The Engineering Department.
    Administrative Working.
    Navigation and River Guards.

SECTION I.

ORIGIN OF PROPERTY RIGHTS AND OWNERSHIP OF STREAMS IN FRANCE.

BASIS OF PROPERTY RIGHTS.*

While under the dominion of Rome all matters pertaining to the streams and waters of the country now called France were subject to governance by Roman law.

Long before the close of the Roman rule, the people had the full protection due citizens of Rome, so that at the time of the conquest of Gaul by the Visigoths (A. D. 470 to 480) there was much land held in individual ownership with the consequent private rights on small streams; but under these Merovingian kings the freehold titles to land disappeared, property was held by a different tenure under the sovereigns, and all right of ownership in water-courses and waters was vested in the rulers themselves.

The feudal system then grew up, and the water-courses, from having belonged, according to their class, to the nation and the people or to

---

*See, particularly, Dalloz and Malapert; also, Dumont and De Passy.

private individuals, under Roman law, and then exclusively to the kings, under Merovingian rule, became dependencies upon the fiefs of the feudal counts, who assumed almost complete ownership of and control over them (9th to 12th centuries).

Actuated by desire for the revenue to be had from tolls and subsidies for navigation and ferry or bridge permits, for several centuries a struggle was now ever present between these nobles and the kings, for the control of the water-courses; and the conflict did not cease until the government had become centralized and feudalism had been overthrown during the fourteenth century.

"All streams and waters belong to the king by right of kingship" was the principle proclaimed by the sovereigns and their nearer adherents.

But in contending for this principle against the nobles and provincial states, the kings in fact gave up control of non-navigable streams—those upon which tolls could not be collected for ferriage and navigation permits—to the bank-land owners.

In the fourteenth century the study of the Roman law was actively revived in France, and the time being about coincident with the decline of feudalism, and the Roman law, recognizing ownership of streams not of public importance—that is non-navigable streams—by the riparian proprietors, this rule apparently thus became incorporated into the law of France.

The kings asserted their ownership of all navigable streams, and those which were floatable for rafts and large timbers, extended the application of the rule as far as there was then any justification for it, and left the control and virtual ownership of non-navigable and non-raftable streams to the bank owners, but really without any formal laws or declarations upon which was grounded their claim of title to them.

The public possessions of the kings, held for the benefit of the nation, became in course of time known as the "public domain," and in 1566 was issued the edict of Moulines, which declared the imprescriptibility and inalienability of this public domain.

This policy of holding fast to all the nation's property, though often attacked, is still adhered to by the government, so that water-courses and waters, once declared navigable and raftable can never be alienated from the public domain, and become in any sense private property.

OWNERSHIP AND CONTROL OF NAVIGABLE STREAMS.*

Navigability and floatability for rafts and large timbers became the

---

* See, particularly, Dumont, pp. 1–14, 135–146, Dalloz, Vol. XIX, pp. 337–337, Proudhon, § 816; also De Passy.

test for streams belonging to the king, but any stream deemed of public importance might have been declared thus navigable or raftable, and made so in sufficient degree to justify its incorporation into the public domain.

The changes in the form of government, occuring a little less than a century ago, appear to have resulted in no completed action affecting the laws or customs respecting waters until 1803-4, when the Code Napoleon was promulgated.

The Code Napoleon is the present civil code of the country. With respect to water-courses and waters it makes this distinct announcement:

"Article 538. Highways, roads and streets at the national charge, rivers and streams which will carry floats, shores, ebb and flow of sea,[*] ports, harbors, roadsteads, and generally all portions of the national territory, which are not susceptible of private proprietorship, are considered as dependencies on the public domain."

This is the only direct statement relating to the ownership of water-courses or waters in this code.

A royal ordinance of 1835 enumerated all the streams and parts of streams in France, deemed navigable or raftable, and hence claimed as of the public domain, and other ordinances, etc., of later dates have added to the list.

The sovereign authority to declare streams navigable, and thereby make them part of the public domain, has not been disputed either in the courts or before the council of state, but riparian proprietors who have been dispossessed of their right to water for irrigation, by the exercise of this power, have claimed, and been allowed by the courts, in a manner prescribed by law, indemnities for actual damage caused them.

Furthermore, although only certain streams and parts of streams, embracing probably all that really are navigable, or that can be made so by a small amount of work, have been thus added to the public domain, the administration, in council of state, may at any time declare other streams or parts of streams navigable or raftable, and thus make them public property, afterwards paying the riparian proprietors for whatever actual damage they may suffer, as may be adjudged by the courts.

The state, owning these water-courses, is, of course, owner of the waters forming them, and these, with the beds, under the edict of

---

[*] These are the words of Richard's translation of "*lais et relais de la mer;*" but the phrase should be rendered "the land left uncovered by and recovered from the sea"—namely, the land newly made by the sea. [Dalloz, Vol. XXXVIII, p. 208.]

1566, are inalienable from the public domain: their *use* only can be granted, as will hereafter be seen.*

OWNERSHIP AND CONTROL OF STREAMS NOT NAVIGABLE NOR RAFTABLE. ¿

The ownership of the beds and waters of streams neither navigable nor floatable for timber, and not claimed as such by the government, is a point which has been much disputed.

It is stated by some French writers of forty years ago that the beds of such streams belong to the riparian proprietors, and they imply that the waters are a sort of property held in common by these proprietors.

But all authoritative writers now hold that "according to the terms of article 714, civil code, water-courses not navigable nor raftable are common property, *i. e.*, enter into the class of things which do not belong to any one."—[De Passy, p. 297.

This article 714 reads as follows: "There are things which belong to no one, and the use whereof is common to all.

"The laws of police regulate the manner of enjoying such things."

But the preceding article, 713, says that "Property which has no owner belongs to the nation."

Taking these two articles together, if the ownership of non-navigable and non-raftable water-courses cannot be fixed elsewhere, then these streams belong to the nation, just as well as do those which have been made part of the public domain by declaration of navigability, under article 538.

The facts are, that riparian proprietors claim the ownership of the channel beds to the center line, each in front of his property, and that the courts allow the claim when the beds are permanently laid dry from any cause; that alluvial deposits along their banks accrue to the benefit of the land owner adjacent to whose field they form; that islands forming in the channels, belong to the adjacent bank owners, in proportion on each side to local circumstances, and that prior to the passage of a law specially to the point in 1847, the owner of one bank could not, even after having secured administrative authority to build a dam in the stream in front of his property, obtain the right to carry it past the center of the stream, or connect with the opposite bank, without the consent of the owner of that bank.

We see, therefore, that until very recent years the beds of streams of this class belonged to and were under the control of the riparian proprietors, except, as will be seen hereafter, in matters wherein the

---
* De Passy, p. 297.
¿ See, particularly, Dumont, B. II, Chs. III and V; De Passy, Ch. 1, and p. 279, *et seq*; De Buffon, Vol. II, Sec. I; Dalloz, Vol. XIX, pp. 379-384; also, Code Napoleon.

government has exercised a supervision of works and channels to insure a free flow for flood waters.

RIPARIAN CLAIMS TO THE WATERS.*

The waters of non-navigable and non-raftable streams were formerly also claimed as the private property of the riparian proprietors. Circumstances of their origin and division, and the necessarily common control of the streams, upset this theory, however, long before the passage of the Code Napoleon.

They were then claimed by these bank-land owners, as a sort of property held in common by them as riparian proprietors, for the exclusive benefit of their lands and industries.

On the other hand, it was and still is claimed by the owners of lands not bordering the streams, that these waters belong to the whole people of France, or are held by the nation for the benefit of the whole people; and while the riparian proprietors are given, by the Napoleonic code, a right to use them in irrigation and otherwise, they are not given an *exclusive* right, but that the government, as the guardian of the waters, can, as in the case of the waters of navigable streams it does, grant concessions for the use of some part of them on lands not riparian, so long as rights already accrued by use, be not unduly or injuriously limited or their exercise inconvenienced by such action.

Replying to this, the riparian proprietors now say, that, if the waters belong to the whole people of France or to the nation, they, the bank owners, have, under the code, a special and complete servitude on all such waters, which servitude, or right to use, is continuous and not forfeitable by failure on their part to avail themselves of it at any time, or for any length of time, except as between themselves, as will hereafter be shown.

The question of the ownership of these waters, and that of the nature of the right to use them which riparian proprietors have, are points of several centuries of litigation in France, for these questions were in dispute long before the civil code was promulgated, and it only changed the aspect of affairs and stirred litigation up again on a slightly different basis, with many fine points of law brought to the front.

The fact of the ownership of the waters of non-navigable and non-raftable streams by the nation, as representing the whole people, is now pretty well settled, and the tendency of decisions and administration rulings, is towards a declaration of ownership, by the nation,

---
* See, particularly, Dumont, Dalloz, and De Passy; also De Buffon; as last cited.

of the beds also, so long as occupied by the waters—or, so long as they are courses for public waters.

Starting several centuries ago, with almost complete ownership and control of the waters and channels of streams not navigable nor raftable, the riparian land owners have since been restricted in their rights, from time to time, and we now find them without any recognized claim of ownership in the waters, and only the semblance of ownership in the channel beds until after these shall have been laid dry; but with a preferred privilege to the use of the waters, as we will hereafter see.

SECTION II.

WATER LAWS AND REGULATIONS.

MOVING CAUSES OF DEVELOPMENT.*

The water laws of France have their roots in the groundwork of principles governing the right of property in water-courses, which have already been spoken of; for the application of these principles, molded by the temper and the wisdom of the rulers, and mellowed by the customs of the people, has brought out the laws and administrative system which we now find.

For centuries in the past, agriculture has been the favorite pursuit of the French, and the rulers of the country have been alive to the importance of fostering it.

Manufacturing, largely dependent on agriculture, has been its branch of industry next in importance.

Its agriculture, in many quarters, has necessitated the application of water in irrigation; its manufacturing has in a high degree been built up by the application of water for power, and has developed a necessity for the use of water in large quantities in very many industrial processes.

The necessity for cheap internal transportation facilities early developed a policy of river improvement and canal construction, so that, commencing in this way as far back as the tenth century, France now has a network of navigable waterways extending over almost the entire valley portion of her territory.

Her hydrographical system and topography is such that large areas of country in her river valleys have been subject to periodical inundations, resulting in loss of property and in unsanitary conditions,

---

* See, particularly, Malapert; also, De Buffon, and many papers, etc., in the *Annalles de Ponts et Chaussées*, and Dalloz, word cited.

producing fever epidemics, thus pressing upon the attention of the people and the government the necessity for improvement of arterial drainage lines, the embanking of lands, and the sanitary drainage of lands embanked and those otherwise subject to receiving too much moisture.

These things all combined, have brought about the making of laws, the growth of customs, and the promulgation of regulative decrees relating to the improvement and guarding of water-courses and waters, and their use in every way and in all interests.

SPECIAL REGARD FOR IRRIGATION.*

Agriculture being a leading interest, and the country ever alive to its importance, we may be sure that it has been fostered, and, indeed, the law-making powers have seemingly ever tried to favor it in the framing of water laws; and although we find much complaint on the part of French hydraulic-agricultural writers, that more has not been done by government in behalf of irrigation, drainage, and the like, and although undoubtedly more could have been done to great advantage, in the way of systematizing matters as well as in the construction of works, yet, considering the political troubles which have for long periods of time disturbed France, on reading the accounts of and laws relating to her hydraulic agriculture, we, who may judge without prejudice, will be led to believe that the French rulers and governments generally have striven to encourage and develop irrigation, drainage, and reclamation, and, in fact, have accomplished much for them. With respect to irrigation—our present subject—we find it constantly favored in the laws, in preference to manufacturing and many other uses of water—domestic necessities and navigation, alone, ranking it in the scale, and the first of these two uses being the only one decidedly preferred to it in the administration of the laws.

In view of these facts we may conclude that the French laws respecting irrigation are about as liberal as they could be made under the circumstances surrounding their development or formation. And it is a significant fact that, although framed for the most part in the midst of monarchial surroundings, amidst all the tearing to pieces which the institutions of France have repeatedly had by the liberalizing spirit that has from time to time prevailed, and even now that the country for a decade has had a republican form of government, the old administrative ordinances, the old administrative system, still prevail almost unchanged, except in details and developments that in no way affect their leading principles.

* See, Dumont, and De Buffon.

CLASSIFICATION OF WATER LAWS.[*]

The earlier laws of those which now exist are the edicts of kings, from the sixteenth to the present century. Then come, in addition to similar promulgations, the decrees of ministerial officers, the enactments of legislative assemblies, the opinions of superior administrative authorities, as well as decisions of courts.

In fact, besides the statutory law, which has grown out of king-made law, and that emanating from legislative officers or bodies, the water laws of France comprise two branches of what answers closely to our common laws in the method of their development—namely: through the interpretation of law and the establishing of precedent by decisions. These branches of the water law have grown, respectively, from the decisions of courts, and from the decisions or rulings of superior officers or bodies of the advisory and executive branches of the administration.

Although on several occasions within the past century, and notably within the last ten years, efforts have been made to bring into form and within a code of small compass the water laws and regulations of France, there is still no general and comprehensive law or code on the subject, but the system is made up of numberless edicts, ordinances, acts, decrees, rulings, decisions, instructions, and circulars, which form a body of law and regulative rules most difficult to trace through in its connections and bearings.

SECTION III.

THE ADMINISTRATION.[§]

Water-courses and waters in France have been from an early period in the modern organized government succeeding the feudal system, generally subjected, not only to laws made by the law-making power of the land and interpreted by the courts, and to regulations made by the executive branch of government, but also to an active and constant supervision by the officers of an administrative organization under this executive bureau.

This branch of the government is called *the administration*. Its regulative measures appear under the titles of decrees, instructions, regulations, administrative laws, etc.

Its purpose, according to the French water-law writers, is to supply the deficiency which must ever exist in the application of general

---

[*] See, Dumont, De Passy, and *Les Annalles des Ponts et Chaussées*, Vols. Laws and Decrees.
[§] See, particularly, De Passy; also, Dumont, and De Buffon.

laws and principles to the management of the affairs of water-courses through the medium of courts: a deficiency which makes itself apparent in the impossibility of fully utilizing streams and waters under any system of primal principles rigidly adhered to under all circumstances throughout a country.

From the difference in the nature of property rights on streams of the two classes—those navigable or raftable and those not so—and from the great difference in the interests to be conserved upon them, result the very essential differences in the administrative policy and measures to which they are subjected.

On non-navigable and non-raftable streams the administration, in theory, interferes with private operations conducted by those who as bank owners have rights on the streams under the ancient usages and the civil code, primarily, to regulate works in the channels or on the banks, with the view of preserving the channels in the interest of the public, and as far as possible assuring or developing a free passage for flood waters without augmenting danger of floods; and, secondarily, with the view of preserving the interests of navigation on the main stream below.

On the water-courses of the public domain—those declared navigable or raftable—the policy of the government is actuated, primarily, by a solicitude for the interests of navigation, and then by an almost equal interest in promoting the economical and full use of the waters in agriculture, manufacturing, and industrial pursuits generally, and, finally, none the less, by a realization of the pressing necessity for promoting the arterial drainage of the country, in order that great floods be prevented, that valuable lands be reclaimed to rich taxable districts, and that insalubrious swamps be reclaimed to healthful neighborhoods.

ADMINISTRATIVE PURPOSE AND POLICY.

On non-navigable and non-raftable water-courses the administration is not authorized to interfere between the owners of works already constructed and those proposed or newly constructed. If a proprietor has lands bordering on the stream, the administration is bound to presume that he has the right to water from it, and it can only interfere in an authoritative way with his project, to the extent of regulating his works, with the views set forth in the second paragraph above.

Further than this, in these cases, the administrative engineers can advise the parties at interest, and bring before them all the facts as to measure of water supply, and extent of use, and nature of existing irrigations, or other data from which to judge of the equities in each case; but if on such showings, amicable agreements can not be arrived at, the administration has no alternative but to sanction the construction of any new work proposed—provided the work itself is unobjectionable—and thus leave the courts to decide, on the showing of facts, whether or not the new appropriator is entitled to water.

On navigable and raftable streams, the administration is invested with full powers, not only to regulate works of all kinds, and much more in detail than on non-navigable and non-raftable streams, but, also, to consider all questions relating to water privileges, to issue and restrict them at will, under the laws.

In the case of both classes of streams, the administrative engineers are charged with the duty of collecting and arranging the data respecting the supply and use of waters, so necessary in an equitable and business-like adjustment of the many questions which arise between the various parties immediately at interest, and between these and the interest of the public, and also so essential to the study of economy and efficiency in use of water, and the full development of the industries dependent on it.

Thus, the administration on non-navigable streams regulates only the works, and the courts adjuge the rights, while on navigable streams the administration adjusts and decides all questions, and issues all privileges, and, finally, on all classes of streams, it obtains the data from which to judge of questions which come up. This much for the scope of power, policy, and duties of the administration as affected by the classification of streams and interests at stake.

GOVERNMENT ORGANIZATION.[*]

France has an area of 204,091 square miles—a territory only about one eighth larger than the State of California.

The country is divided into 87 departments, these into 362 arrondissements, or sub-departments, these into 2,863 cantons, or judicial districts, and these, finally, into 36,056 communes, or municipalities.

France is a republic, but very many of her institutions are monarchial by origin and in spirit. The legislative power is vested in two houses, or chambers—the chamber of deputies, and the senate; and the executive authority, in a president.

The chamber of deputies is elected by universal suffrage, each arrondissement being represented by one deputy, or by more if its population exceeds 100,000 souls. The senate is composed of 300 members, of whom one fourth are elected by the senate itself, for life,

---

[*] See, particularly, Reclus, Vol. II, Ch. XV; also, Malapert, Ch. XXI.

and three fourths are elected for nine years by electoral colleges formed in every department and commune.

The president is elected by the senate and the chamber, sitting conjointly, for seven years. The president promulgates the laws voted by the chamber; and he appoints his ministers, who are responsible to the chambers for the conduct of their several bureaus.

A council of state, presided over by the minister of justice, and consisting of thirty-seven councilors and twenty-four masters of requests, nominated by the president, and thirty auditors, nominated concurrently with the senate, advises on laws referred to it by the chambers or by the ministers, and on all matters submitted by the president, performing in this way certain duties as the chief advisory and regulative body to the bureau which has to do with the administration of the affairs of water-courses and waters.

Each department has its general council, the members of which (generally one for each canton) are elected by universal suffrage, for six years. These councils meet annually to discuss the department budget and to act as advisors of the prefect. The prefect is appointed by the president, on nomination by the minister of the interior. He is virtually the governor of the department, and his powers are extensive.

Each arrondissement, or sub-department, has its sub-prefect, and a council elected by universal suffrage, to consider and regulate purely local matters. The cantons are merely judicial and electoral districts.

Each commune has a municipal council of from twelve to eighty members, elected by universal suffrage. The mayor of the commune is appointed by government, but he must be a member of the elected municipal council. He represents the state as well as the commune.

THE ADMINISTRATIVE SYSTEM.*

As will be seen, the mayors and municipal councils, the sub-prefects and sub-prefectorial councils, the prefects and the prefectorial or general councils of the departments, as well as the council of state of the government, are all connected with, and in fact together, make up the administrative department, which, with the engineering department and bureau of public works, control the affairs of watercourses and inland waters of the country.

The mayors and the prefects are the principal administrative units in this administrative system, and to give an idea of the scope of their territorial authority, it may be remarked that the average

---
* See, De Passy, Malapert, and Reclus.

commune is 5.5 square miles, and the average department is 2,345 square miles in area.

This makes the jurisdiction of a mayor cover territory less than one sixth of a township of our land survey system, and shows the average scope of country presided over by prefects to be about the size of Colusa, or one half that of Los Angeles county, in this state.

THE BUREAU OF PUBLIC WORKS.[*]

The construction and management of all public works, except those specially and fittingly confided to the minister of war, of the navy, of education, of posts and telegraphs, and some others, is delegated to the secretary of state or minister, of public works.

Amongst the duties confided to this authority are all relating to the hydraulic service; to ports, harbors, coasts, rivers, streams, canals, torrents, irrigation, drainage, reclamation, and the like.

The care of all waters and water-courses, whether of the public domain or not, their control, and the control of the acts of individuals on their banks, is regarded as of public concern, and the administration has to do with the affairs of all streams, in a greater or less degree, as will hereafter be seen.

The minister of public works is the chief executive officer of government in this branch of the organization. He acts under authority of laws of the country, and in the light of opinions or interpretations of old laws and customs, by the council of state.

He himself makes rulings and regulations in conformity with principles thus laid down, in his circulars and instructions to subordinates.

For this purpose of administration, the prefects, each in his department, are the chief local executive officers under the minister of public works.

In the management of the affairs of the streams, in all, except the planning and superintendence of work, and the experting of all questions of a technical nature connected with the subject, the prefects act under authority, and in accordance with the ministerial circulars and instructions, which communicate to them the results of, or the texts of, the advices of the council of state, when such there be.

Thus, all applications for permits or authorizations, or executive rulings, or enforcement of regulations, first come to the prefects, and they, if endowed with the authority suited to the case, act on it, or refer it with comments to the minister of public works, if not competent to decide themselves.

---

[*] See, particularly, Malapert, Ch. XXI and elsewhere; also, De Passy.

THE ADMINISTRATIVE SYSTEM.

THE ENGINEERING DEPARTMENT.*

In the ministry of public works is a bureau of civil engineering, known, from long ago, for reasons not necessary here to explain, as the department of *bridges and highways*.

This bureau is a very extensive organization of men scientifically and practically educated at a government school for the purpose. Their mission is civil engineering, primarily, and not military engineering or the art of war. The organization is somewhat that of the officers of an army, but promotion is not altogether by seniority, for competency and special fitness have much to do with this.

From this bureau, engineers are detached to other service—to the department of war, to that of agriculture, to that of posts and telegraphs, to the service of cities, and on special works, etc.

The greater portion of the engineers of bridges and highways are in the immediate service of the ministry, or department, of public works, in the construction or management of public works, or the supervision of private works or operations affecting the public domain, or the common welfare of all the people. While others of these engineers are in the service of the departments, and more directly charged with advising the prefects and prefectorial councils.

Wherever they go, however, their plans of works proposed are subject to revision by the central commissions of the corps, and all technical matters of great importance are referred to the engineer-in-chief, to be by him laid before the proper revising board.

Besides the engineers, there is a corps of "conductors," who are the superintendents of works. These men, besides a certain theoretical training, have a practical education as constructors—stone and brick masons, carpenters, and builders of all kinds—and each one is a master in certain branches of practical construction.

They report to the engineers, and carry out their plans and specifications. The conductors are graded, and have various ranks in their corps; and after a certain service become also advisors and inspectors of works.

The engineers are the executive officers of the minister of public works, in carrying out all works of a distinctively public character, and also in the preliminary examinations for, supervision of, and reporting on all private or other works affecting the public domain or the common good. And they are the advisers of the prefects in the regulation of matters pertaining to waters and water-courses, as well as other things.

---
* See, particularly, Malapert, Ch. XXI, and the heading "Engineers" in preceding chapters; also, De Passy, supplement.

The management of works of navigation, such as locks, dams, etc., on canalized or improved rivers, and of public canals of navigation, and of works for the diversion of waters from streams, is intrusted to their charge.

In a measure they have a co-jurisdiction with the prefects in some matters of police of streams, and the line of duty of each is the subject of careful designation by ministerial decrees and instructions.

Of the duty and authority of engineers and prefects more will be seen in the chapters which follow.

ADMINISTRATIVE WORKING.*

Briefly reviewing that which has been said applicable under this heading, we see that the administration of waters and water-courses is confided to the minister of public works and his subordinates of the engineering and executive corps in the hydraulic service, and to the prefects of the departments, who, acting independently in some things, are still wholly accountable to the minister in others.

Thus, in matters pertaining to the construction of any particular or important work, or the granting of any water privilege on navigable streams, the prefects can only act provisionally, and every case has to be considered by the minister of public works, and advised upon by the council of state.

In matters of simply carrying out resolutions and the minor works of repairs or construction on this class of streams, the prefects have authority to act without reference to the central administration, but an appeal may always be taken by parties at interest to the minister or council of state, from an order or ruling of a prefect.

On streams not of the public domain, prefects have authority to grant privileges for the construction of all works, when they are duly advised by the engineers that no harm will be done by them and that the plans are commensurate with the purpose in view.

And so, on this class of streams, the prefects are intrusted with the administration of all regulations, and the making of regulations for matters of detail in carrying out the decrees of the central administration and the decisions of the courts.

The prefects of the departments, in performing executive duties, act through the sub-prefects of the arrondissements composing their departments, these through the mayors of the communes composing their arrondissements, and these through the river-guards and rural police of the country.

The government civil engineers form almost a distinct line of

---

* See, particularly, De Passy: also, Dumont, and Malapert.

executive officers, as directly accountable to the central administration as are the prefects.

Those who are assigned to duty as departmental engineers are, of course, annexed, as it were, to the staff of the prefect in each instance, but those not thus assigned are in no way accountable to the prefects, except as they may be placed to advise on works or measures with which the prefects may have to do.

The navigable streams and navigation canals of the country are under the supervision of engineers, the duty being apportioned so that one engineer is in general charge of a whole work or system, and all others connected therewith are accountable to him.

The engineers on this duty act through their local assistants, and these through the guards of navigation, hereafter to be spoken of.

The departmental engineers have to do more particularly with the non-navigable streams, and are in this line of duty the advisers of the prefects, and, in the absence of engineers specially in charge of any navigable stream, the departmental engineer is the adviser.

Thus, on navigable streams, in matters of management and maintenance, the engineers are really the executive officers and the advisers of the central administration, while on non-navigable streams the prefects are the executive officers, and the departmental engineers advise them. In matters of permits and privileges, the prefects are the executive officers on both classes of streams, and the engineers the advisers. In matters of construction, the engineers have exclusive control on navigable streams, and are the supervisory officers of private works on non-navigable streams.

NAVIGATION AND RIVER-GUARDS.*

It is now the intention of the government, that all water-courses of public importance in France, whether navigable or raftable—and, consequently, of the public domain—or not floatable even for rafts, and timber, but which—by reason of the use of their waters in irrigation, or for power, industrial, municipal, or other purposes, or by reason of the existence of levees on their banks, or of their channels being outfall drains for populous districts, or by reason of their being tributaries to navigable streams where water supply is scarce—are of public utility, or liable to receive injury to their channels or banks, or to do injury by the excess or failure of their waters, shall be subject to the supervising care of special agents of the government, called *guards*.

On navigable streams these agents are called "guards of navigation;" are appointed by the administrative officers in general charge

---

* See, particularly, De Passy, and De Buffon, pp. 98 to 106 and elsewhere; also, Malapert.

of the construction, maintenance, and operation of the works of navigation, under the direction of the minister of public works, and are paid by government.

The "guards of navigation" have charge of the operation of all locks, movable dams, sluices, and other structures in the river channels, and of all gates, sluices, or other openings for diverting water through the banks. They in fact perform the duties on rivers, similar to those performed by a superintendent and his assistants on a canal in his charge.

At the principal structures, such as locks and movable dams, guards are of necessity stationed all the time, while others are assigned to beats on the river along the intermediate reaches.

Every river being subject to general regulations laid down by the central administration, and to special regulations covering details and laid down by the local administrative and engineering authorities, it is the province of the guards to see that these are observed and not infringed upon; to see that all who have water privileges get their dues according to the schedule, and are not curtailed in their enjoyment of them by the greed, carelessness, or ill-feeling of others; to see that, neither by neglect nor criminal act, is anything done to injure the bed, channel, or banks of the streams; to observe all works connected with navigation or affecting the stream in any way injuriously, and to report their condition; to prevent the deposit of filth, rubbish, or dirt in the channel or on its banks; to keep a record of the flow of the waters, and of their height at different points; and also to render assistance, in cases of necessity, to river-craft crews or others endangered or embarrassed from any cause.

Some rivers are specially under the charge of engineers detailed from the government civil engineering corps for the duty; and in these cases the guards report to them and receive instructions from them. In cases where the navigation is not thus exclusively under engineering control, the guards are subordinated to some other government functionary having these interests in charge, perhaps in the several localities.

RIVER GUARDS—THEIR DUTIES AND COMPENSATION.*

On non-navigable streams, the guards are called "river-guards." They are appointed by the prefect of the province, generally on the recommendation of the riparian owners, and others interested on the stream, and are paid by the prefect, with moneys collected from the parties at interest on the stream, according to circumstances.

---

* References same as those for preceding heading.

On streams where waters are used largely for power purposes, and which are not embanked, or, from other cause, threaten riparian lands, the tax for the salary of the river-guards is levied entirely on the manufacturing interest using the water, or, if at all, in a small degree, only, on the owners of riparian lands.

While on streams embanked, and threatening overflow of adjacent lands, and on streams used as drains for riparian properties to a considerable extent, in the absence of manufacturing interests, the salaries of the guards are assessed wholly on the riparian proprietors.

Still again, on streams whose waters are used in irrigation, to the exclusion of other uses, and where there is no special reason, as first above mentioned, for calling on riparian proprietors not thus using water, the salaries of the guards are assessed chiefly upon those who divert the water, and the riparian proprietors not diverting water, pay but a small portion.

These rulings are the outgrowth of custom, and while they are very generally accepted without opposition, they have met, and still do meet, in some cases and localities, with very strong opposition from those who have to pay.

On these non-navigable nor raftable streams, the river-guard is a supervisor or inspector of maintenance of works, and a police inspector to report the condition of the streams, banks, and channels, and to report all acts of omission or commission in contravention of the general laws and special regulations applicable to the river, or part of river, placed in his charge. He is assigned a regular beat, over which he has to go at stated intervals, examining everything pertinent to his charge, keeping a minutely detailed journal of his operations, and reporting to various officers, designated in each case, at different parts of his district. The following is a formula for duty for river-guards used in the regulations in the department of *Seine-et-Oise:*

"A river-guard is specially charged with seeing that the present regulation is observed; that the execution of the works of cleansing the channels, remodelling and protecting the banks, cutting away undergrowth where harmful, mowing the tall grass or rushes on the banks, etc., are carried out according to the orders of the syndicate and of the engineer of the district, and under the surveillance of the mayors of the communes traversed by the water-course, that is the subject of this regulation.

"The river-guard must report all infringements whatsoever of regulations committed by manufacturers, riparian owners, or any other person. He must visit once a week all parts of the river intrusted to his superintendence, and prove the fulfillment of his duty by the signature of the local officers in the various parts of his district, to whom he reports.

"He must keep a daily register numbered and indexed by some

proper superior officer, in which he inscribes every day a report of all the facts that come to his knowledge on his tour of inspection, and particularly all infringements of regulations, or offenses that come under his observation.

"Once a week at least he reports to the chief officer of the district, to whom he is accountable, or to some other specially delegated authority, to give a verbal account of all that he has seen, to have his register examined and countersigned."—[De Buffon, Vol. II, p. 102.

THE NECESSITY FOR RIVER-GUARDS.*

The necessity for river-guards is generally dwelt upon by writers on the subject of water-courses in France.

De Buffon, perhaps the most authoritative author on the general subject of hydraulic agriculture and the management of water-courses in various countries, that has ever written—says on this point:

"Every day experience shows that the operations necessary for the preservation of stream channels, would soon be without useful results if they were not kept under strict surveillance by agents beyond the power of local control. Worse than all, the works of maintenance and repairs of structures so necessary to insure security of property from overflow would not be executed were the proprietors not closely watched.

"According to this double motive, wherever the utility of these works has become well understood, those interested have recognized that the influence of a special general agent is indispensable to insure the measures of construction and police in question."

And in another place this author says:

"This principle is admitted by every one who in the least understands the matter, that water-courses not a part of the public domain are, in the absence of governmental control, really in a state of abandonment, that seems to call forth on the part of riparian owners manifold offenses against the common welfare, and usurpations of all kinds."

"The first consequence of this state of things, deplored by everybody, is the enormous damage thus caused to agriculture, increasing daily, and so occasioning losses whose amount in coin, if it could be calculated, would be a frightful sum; lessening the agricultural wealth of the country, wherever this interest in the streams has for any length of time been neglected."—[De Buffon, Vol. II, p. 133.

This damage is depicted as arising from injury to stream channels by want of care and neglect, or the deposit of materials so as to cause the filling of the channels over long courses, and the consequent overflow of lands, or the supersaturation of soils with moisture from the effects of bad drainage.

And the author then says:

"We could cite localities, rich and flourishing in years gone by,

---

*See, De Passy, Malapert, Dumont, and De Buffon.

where to-day agriculture is nearly annihilated, under the weight of calamities which were preventable by proper guarding of the small streams. Far are we from exaggerating the real situation to attract attention to the subject we are occupied with, for it is easy to assure everybody of its truth, or, to say it better, it is a truth too well known, for everybody can prove it by investigations in many localities, by the weight of the mournful words: 'We average a crop in but two years out of five.'"—[De Buffon, Vol. II, p. 134.

"The indispensability of river-guards must be considered as having been completely demonstrated by experience. The practice of riparian owners and manufacturers making encroachments on the channels of water-courses has, in every instance, developed where there has been no inspection, or where the agents had too extended beats and could not attend properly to their duties. But where guards have been in almost daily communication with the users of the water-courses, regulations have been observed and infringements prevented. An infringement taken at the commencement is generally discontinued on receiving a friendly notice, while suits entered afterwards are often uncertain in their results."

These words were written in 1856, when the hydraulic service of the country was not nearly as well organized or extended, nor the regulations so strictly enforced as they are now; and it is considered that de Buffon contributed more than any other person to the general understanding and popular appreciation of the subject at large, and thus did much to forward measures of reform which have since followed.

The sentiment actuating these measures, and the principle on which they rest, are aptly set forth by the following paragraph from the same work, written in the discussion of the habit of encroachment upon and interference with stream channels and banks by riparian land owners:

"Water has, on riparian properties, a natural, primordial right— the right to a sufficient and proper channel in which to pass. * * * River waters are, then, from time immemorial, in possession of canals carved out of the surface of the earth, in dimensions proportioned to the quantity of the flow to be carried. This is possession on the part of the state. The existence of these canals, as old as the world, is a title in the state, inscribed in the ground by the hand of God for the common good. Consequently it is a sound conclusion that public authority should have the right, and that it should be its duty, to have them respected and not tampered with by every dweller on their banks."—[De Buffon, Vol. II, p. 148, etc.

AUTHORITIES FOR CHAPTER II.

In the preparation of this chapter I have consulted the following named authorities:

De Passy.—"A Treatise on the Hydraulic Service." By M. G. De Passy, a chief engineer in the Government Corps of Civil Engineers, France; 1 vol.; 3d ed.; Paris, 1876. See, generally, pp. 1 to 130, 297 to 385.

Dumont.—"The Legal Organization of Water-Courses." By Messrs. Adrien Dumont, Advocate, etc., and A. Dumont, Corps of Civil Engineers, France; 1 vol.; Paris, 1845. See generally, pp. 1 to 13, 134 to 330.

De Buffon.—"A Course of Agriculture and Agricultural Hydraulics." By Nadault de Buffon, an engineer-in-chief of the Government Corps of Civil Engineers, France; 3 vols.; Paris, 1856. See Vol. II, Part II, Section 1.

Malapert.—"History of the Legislation of the Public Works of France." By W. F. Malapert, Advocate, etc.; Paris, 1880. See, generally, headings "Water-Courses," "Engineers," "The Present Republic."

Reclus.—"The Earth and its Inhabitants." By Élisée Reclus; edited by E. G. Ravenstein, F.R.G.S.; 5 vols.; New York, 1881. See Vol. II, France, Ch. XV.

Dalloz.—"Methodical Treatise on Legislation and Jurisprudence." By M. D. Dalloz, Sr., and M. A. Dalloz; 56 vols.; quarto, Paris. See word "Waters," Vol. XIX, pp. 312 to 500.

Debauve.—"Waters as a Means of Transportation—River Navigation." Vol. XIX of authorized Civil Engineering Manuals. By A. Debauve, Government Corps of Civil Engineers, France; Paris, 1878.

Debauve.—"Use of Waters in Agriculture—Irrigation," etc. Vol. XVIII (of above named series).

Proudhon.—"Treatise on Ownership of Property." By M. E. Proudhon; Bruxelles, 1842. See ¿ 815-820.

Civil Code.—French Codes. Edition of Alphonse Pigoreau, Paris, 1845.

Civil Code.—Code Napoleon. Translated by Robert S. Richards, M.A., London.

CHAPTER III.—FRANCE[2];

WATER PRIVILEGES AND THE ADMINISTRATION OF NAVIGABLE AND RAFTABLE STREAMS.

SECTION I.—*Water Privileges.*
 The Uses to which Water is Put, and the Regulation of its Use.
 The Rivers and River Works of France.
 Forms of Organization of Enterprise.
 Applications and Formalities for Water Privileges.
 The case of the Bourne Canal.
 Obligations of the Grantees.
 Conditions of the Concession.
 Privileges of the Grantees.
 Benefits to the Grantees.

SECTION II.—*Regulation of Works.*
 Government Improvement of Rivers Generally.
 Extent and Field of the Hydraulic Service.
 The Principles of Coöperation and Compulsion.
 Construction of Dams and of Headworks.

SECTION III.—*Operation and Maintenance.*
 General Maintenance of Works.
 Cleaning or Dredging of Channels.
 Police of Streams.
 Water Privilege Rents.

SECTION I.

WATER PRIVILEGES.

THE USES TO WHICH WATER IS PUT, AND THE REGULATION OF ITS USE.

Water is extensively used from streams in France for irrigation, the production of power for manufacturing, for consumption in industrial processes, for domestic, sanitary, and other municipal purposes; and these four uses will be referred to herein as "irrigation," "manufacturing," "industrial works," and "municipal uses."

As opposed to these industries and necessities which generally require the water to be taken from the streams and in great part not returned, the interest of navigation, the general sanitary condition of the stream channels and consequent healthfulness of their neighborhoods, the convenience, comfort, and sometimes the necessities of riparian land owners, and the gratification of the people generally, demands that the water be left in the streams.

And while thus there is a serious clashing between the two sets of opposing interests, those who demand the water out of the streams are by no means in harmony, but amongst themselves are most often brought face to face by conflicts of interest.

The government owns, controls, and in a business like way, administers the affairs of all water-courses deemed navigable or floatable for rafts, or large timber, fostering the interests dependent on the use or presence of the water, and striving to insure the most complete, widespread, and well distributed good results to the people and the nation from the use of their properties.

To this end, these streams have been studied so that their channels are well mapped out, their flow at different seasons of the year known, the requirements of the various industries well understood, and every work affecting the river's flow, or intended for drawing water from it, is planned and registered, and its rights or necessities, understood.

There are very many old water-rights on these streams dating back several centuries; some even previous to the issuing of the edict of Moulines in 1566, but even towards these the administration has power to act as may best conserve the interests of the public and preserve the equities which attach to the private interests involved.

THE OBJECT OF ADMINISTRATION.*

Interference is not the object of this systemization, nor is it the practice to needlessly exercise surveillance or management of the use of water. The object is to protect each general and individual interest against the general and naturally unavoidable antagonism of each other interest, and to administer a common property, which, by the nature of things, could not by any possibility be administered in a business like way by any other than a governing power of some kind.

Accordingly, no work of any sort, kind, or description may be erected upon a navigable river or a stream floatable for rafts, or timber, or one declared so, in France (indeed, the rule in this regard is not much less strict for streams not navigable or raftable, as well), nor can any water be taken from such streams, except it be taken in a bucket or other similar hand vessel, without the project for which it is required, the plan by which it is to be constructed, if a work, or used, if a

---

* See, De Passy, Dumont, and De Buffon.

water privilege, has been first submitted to the administrative authorities, and publicly made known, criticised, and opposed if necessary.

All interests are put on their guard, all sayings in opposition are heard, all criticisms listened to. The project is examined by those knowing well the facts bearing on the whole case, and competent to judge of the tendency of such facts and their probable results; and permits are issued or refused after the whole case has been viewed with all the care and intelligent consideration which its importance will justify in each instance.

Older rights and those of industries most needful are always protected in the administration of affairs from day to day; but no right is so old or no use so pressing that its owners have the power to control the division of the people's water, or use it in a manner wasteful or inefficient, or in any way unnecessarily hamper or hinder the full development and prosperity of other institutions dependent on water supply.

This is the object and purpose of the French administration of waters It cannot be claimed to be perfect, either in theory or practice. That it is the best devised and in use, befitting application amongst a free and enlightened people, there can be no doubt. But it is not the best that can be devised for freer, equally enlightened, and more progressive people. Nevertheless its main principles are to be noted, and the general ideas of settled and registered privileges, and intelligent administration of the element common to their beneficial exercise, is to be kept in view and incorporated in any system which will assure freedom from clashing, immunity from litigation, and a full measure of benefit from the opportunities presented.

RIVERS AND RIVER WORKS IN FRANCE.*

France has a very much extended and intricate system of watercourses, several of which are large rivers naturally navigable for long distances from the sea.

Rising amidst the snows and glaciers of the high Alps, or on the rain drenched face of the Pyrenees, or in the forest covered and heavily watered Vosges, or upon the rolling and wooded plateau of central France, these rivers are generally well supplied with water, and are sometimes subject to great and devastating floods.

The destructive operations of man and his grazing animals on the mountains, the industrious tillage and soil loosening on the rolling grounds, the wasteful and criminally stupid action of municipalities in the disposal of filth by depositing in river channels, and others of

---
*See, Reclus, Debauve (Vol. XIX), Malapert, and De Passy.

the nature-consuming influences which have unfortunately accompanied the development of civilization, long ago forced the attention of the French government to river maintenance and improvement as a national necessity. So that river works, commenced as purely commercial ventures and enterprises by private individuals and companies and by the government in the centuries that have passed, have been added to in great number and spread out in class and character and locality, over nearly the whole country, by the influence of necessity in preventing harm, as well as that of enterprise in promoting the development of the country.

The lower and larger rivers and those of light slope in alluvial formations have, as a class, been improved by systematic embanking, training of currents, and dredging, and the higher rivers of greater grade slopes, been made navigable by dams in series, retaining the waters, at times of ordinary and low supply, in approximately level reaches from one to the other, or lessening the grades at those parts of their courses where the natural slopes of the beds were too great to admit of a navigable depth with the supply at command, and with a moderate current in the waters.

These succeeding reaches or levels are, of course, connected at the dams by means of water locks for the passage of boats, and the dams themselves in very many instances are partly removable along their crests, sometimes automatically by the rising waters, and sometimes by the work of attendants, so as the better to admit of free passage for flood waters.

In the cases of the higher streams, or parts of streams, the channels are frequently made floatable for timber and lumber passing from the forests on their head waters, also by means of dams, having permanent ways or removable weirs through or over which to float the rafts.

NAVIGABLE AND NON-NAVIGABLE RIVERS.[*]

It is on the rivers and portions of rivers where it has become necessary to construct dams for navigation, and those, still higher, which have been dammed for purposes of floatation, that water privileges are chiefly sought after for power purposes, irrigation, municipal supply, and industrial use.

Such water-courses are public property, under full control of the administration.

Non-navigable and non-raftable tributaries of navigable or raftable streams, and these streams themselves above the points where they become navigable or raftable, for the reason that it is necessary in the

---

[*] See, particularly, Dumont (§ 88 to 91); also, Debauve, and De Passy.

interest of navigation, public water supply, equity in distribution of waters to claimants below, and other reasons obvious from what has already been written, are also under the control of the administration, which is authorized to limit all diversions proportionately, or prohibit them at times, according to prefixed schedules of right, and rules and regulations framed for each case, when necessary to the public welfare.

With respect to the non-navigable arms of those streams which divide into two or more branches in their onward course, the governing rule appears to be not so well defined.

When such non-navigable arm again unites with the navigable channel, it is regarded as being itself navigable, and is subjected to regulations accordingly.

When such non-navigable branches do not again join the main or navigable stream below, according to some authorities, they are regarded as navigable, and the reverse is true, as stated by other writers.

Still or stagnant waters, those draining from marshes and ditches, that have free communication from navigable or raftable streams and whose waters flow the year round, or waters where ferry boats can enter at all times, and those cared for at the expense of the state, also make part of the public domain, and a right to dispose of or use them may be had only by special authorization, as in the case of navigable streams.

FORMS OF ORGANIZATION OF IRRIGATION ENTERPRISE.[*]

Setting aside that very large class of cases brought up by reclamation, embankment, drainage, municipal improvement, sanitary regulation, and other developments requiring the construction of works in or on the banks of water-courses, and which, equally with the class of cases herein to be considered, come under the supervision of the administration, but which are not so intimately connected with irrigation works and the use of water from streams as to justify their treatment in this report, we come now to a glance at the forms which irrigation enterprise takes, and then to the various proceedings made necessary by these varied forms of organization, to acquire the privileges desired by each.

Projects requiring special privileges to use water, or sanction of plans to erect works in water-courses, are undertaken either as private enterprises of individuals to water their own lands, to run their own mills, or for other private purposes, as coöperative enterprises of associated land-holders for the watering of *their* own lands, etc., or as speculative enterprises by individuals, associated land-holders or

---
[*] See, particularly, Dumont, and De Passy; also, De Buffon.

capitalized incorporated companies desiring to sell water to consumers.

These differences of organization, together with the variation in use to which waters are put under the privileges, as already explained, make necessary different forms of application, varied formalities in the consideration of them, and distinctive forms and conditions attached to the grants which result.

Instances of individual enterprise are common on streams of all classes, but most frequent on non-navigable streams, and enterprises in which there are several copartners, rank with those individual.

Associations of land-holders for irrigation usually take the form of "syndicates"—a species of organization provided for by a special law, hereafter to be spoken of (Chap. VII)—and enterprises carried on by these associations are also most common on non-floatable streams.

Speculative enterprises are generally on comparatively large scales, conducted by capitalized companies, and under special grants of water privileges on the larger streams of the public domain.

In order to divide the subject well, and give a range of illustration without taking too many examples, the forms, etc., for individual enterprises, will be spoken of for both navigable and non-navigable streams, the forms for grantee companies, under the head of navigable streams, where alone they could be placed, and the forms for syndicate associations, under the head of non-navigable streams, on which they are most common.

APPLICATIONS AND FORMALITIES FOR WATER PRIVILEGES.*

When water privileges or permits to construct works are desired by individuals for their own private benefit, in the use of water or otherwise, on navigable or raftable streams, a formal application must be made to the prefect of the department wherein the intended work or diversion is to be made.

Accompanying this application there must be a statement as to the object for which the work is intended, and the location, character, and general plan of the work itself. If the application is also for a privilege of using or diverting water, in addition to the specifications concerning the works intended or desired, there must be a statement concerning the use to which the water is to be put, the lands to be irrigated, if any, the amount desired, etc.

Under the direction of the prefect the project is reported on, pre-

---

* See, particularly, De Passy, and many Decrees, etc., in the vols. of the *Annalles des Ponts et Chaussees;* also, Dumont.

liminarily, by the mayor of each commune in which the proposed work is situated, or where its effects will be directly felt. These preliminary reports are made after due advertisement and inquiry, and the hearing of objections on the part of those who may care to oppose the measure. The sub-prefect of the arrondissement, to whom these reports are made, reviews them as he may see fit, and transmits them, with all the papers and abstracts of evidence, to the prefect.

This preliminary examination is made with the view of calling out and collecting the sentiment of the people interested, and as a basis for the other investigations which follow.

The results of the preliminary examination are handed by the prefect to the departmental engineer, or if there is an engineer specially in charge of the stream in question, they are handed to him, with instructions to examine and report.

This engineer then holds an inquiry into the case, with the view of ascertaining the engineering bearing of the works proposed, and the manner in which other works, rights, or interests may be affected, and the public utility of the stream subserved. The engineer may take evidence of interested parties should he see fit, and must always examine the ground and locality of the proposed works.

He draws a report in writing, which is transmitted, with all the papers, etc., to the chief of the engineering bureau and also to the prefect. If it is a case in which the prefect has authority to act, he goes on with it; if not, he awaits the opinion of the engineer-in-chief.

On the basis of the engineer's report the prefect instructs the sub-prefect to hold the final inquiry, notices of which are duly published.

All of the papers are opened to inspection, and the plans to criticism, at the mayoralty house of the local commune.

The engineer may be called upon to revise the plans or to modify the project to suit the case or do away with objections.

Finally, the sub-prefect reports the results to the prefect, and, if it is a case in which his authority is competent, he issues or denies the desired permit or privilege; or if his authority does not meet the case, he refers it to the central administration, which in due time acts by decree of the council of state.

WATER PRIVILEGE GRANTS—EXAMINATION OF PROJECTS.[*]

Where water privileges on streams navigable and of the public domain are desired by individuals, companies, or societies, for speculative purposes, all permits and concessions have to be acquired by decree deliberated upon in the council of state.

---

[*] Same authorities as for preceding subdivision.

In this class of cases a still more formal line of proceedings has to be followed out than those already described for individuals obtaining permits to water their own lands, or for other purposes of private use.

The application for the grant of privileges, etc., is made to the prefect of the province.

It must be accompanied by:

(1) An outline map of the proposed district to be irrigated, showing property divisions and other features, and indicating by special tinting the irrigable lands under the project.

(2) A statement in detail of the extent of each district, with the names and residences of all land proprietors therein.

(3) A statement of the conditions proposed to be attached to the contract of the grant and accepted by the petitioner.

(4) Preliminary plans, specifications, and estimates of the works, drawn out in considerable detail.

The project is submitted by the prefect to the proper engineer, who gives an opinion as to the public utility of and necessity for the works.

The prefect then indorses his own views in this regard upon the report and forwards it to the minister of public works.

Following this and on instructions from the minister, the proceedings heretofore described, in which all interested parties have their hearing, are had under the supervision and conduct of the departmental administrative officers and engineers.

Upon the results of these inquiries being returned to the minister, together with the reports of the engineers, he brings the whole subject before the council of state, with his opinion and recommendation.

Should the petition be acted upon favorably, the minister of public works enters into a contract with the grantee, in such way as to guard the interest of the public and of the landholders in the district, and a decree is issued granting the privileges desired, and stipulating the conditions attached.

Large works of this kind are considered of such great public value in France, and local financial conditions are so much against their undertaking, that the government, as elsewhere explained, on proper showings being made, engages to pay a subsidy to the grantee company, or individual, as the works are carried out and completed.

THE CASE OF THE BOURNE CANAL.[*]

I take, as an instance of such a work and grant, the case of the

---

[*] See, the *Annales des Ponts et Chaussées*, Laws and Decrees, Vol. CXXVI, p. 451, *et seq.*; also, De Passy, p. 363, *et seq.*

canal of the Bourne River, in the department of the Drôme, which was authorized in February, 1874.

Application was made by three individuals, on behalf of a society organized in the region of the proposed irrigation; not as a syndicate of land-holders to irrigate their own lands, but as a company to carry out a project as a business proposition, and to deal with several syndicates of land-holders desiring irrigation for their lands.

The formalities being gone through with, the minister of public works entered into a preliminary convention or agreement with the society, in which terms of the concession were drawn out in detail.

There being some doubt as to the proper proceedings, and a large subsidy being asked, the matter was brought before the national assembly for confirmation by a special law.

This was passed in May, 1874, declaring the public utility of the work, sanctioning the terms of the preliminary agreement made between the minister and the grantees, ratifying the engagement to pay the subsidy, and prescribing a form for the final contract, covering the principal points of the preliminary agreement.

I have made an analysis of these documents, grouping their important points under suitable headings, and here present the results, as follows:

OBLIGATIONS OF THE GRANTEES.

The company is obliged:

1. To build at its expense, risk, and peril, the principal canal, the two additional diversion canals, the secondary canals, and the tertiary canals and ditches intended to lead water to each irrigation proprietor's distributing gate.

2. To maintain the principal and the two diversion canals at its own expense and under its own immediate care, and to maintain the secondary and tertiary canals and ditches, etc., at its own expense and under its own care, or, by an arrangement for the purpose, under the care of the irrigators.

3. To construct at its expense, delivery and distribution works, for water for domestic purposes, for each commune, with branch pipes and faucets to each house entrance for each subscriber.

4. To maintain these works and all parts of them, down to pipes which carry two decilitres (c. ft., 0.007) of water per second.

5. To submit for the approval of the minister of public works, within the year following the giving of the concession, a detailed plan of the dam and the head works to be constructed for the principal canal and the two subsidiary canals of diversion.

6. To completely finish the principal canal from the Bourne River,

in working order, within five years from the date of approval of the concession.

7. To construct the secondary and tertiary canals and ditches for distribution, in each instance, as soon as the subscriptions for water to be delivered by the particular work are sufficient to assure a revenue of six per cent on the cost of the work, according to estimates to be approved by government engineers.

8. To complete each distributing system, when once commenced, within two years.

9. To commence the subsidiary diversion canals—from the Lyonne and Cholet rivers—so soon as water is subscribed for to the extent of five thousand litres (176 cubic feet) per second, and after the commencement of the main canal, and to finish them within two years after commencement.

10. To reëstablish and maintain at its own expense the free flow of all drainage waters, whose course may be intersected by the works.

11. To do all possible at all times according to the rules laid down by the administration to stop seepage waters from the canals and other works built by the company, and stop all undesired wetting of lands and property.

12. To construct, at its own expense, permanent bridges for crossings of all existing ways of communication, encountered by the canal, according to approved plans, and of dimensions specified in the agreement for roads, etc., of different classes and kinds.

13. To construct for use, pending the completion of these permanent crossings and the canal, adequate and safe temporary crossings and side roads for the traffic, according to approved plans.

14. To manage its work according to approved plans so as never to interrupt traffic on any railroad or other principal line of travel.

15. To conform to all rules hereafter made by the administration relative to the preservation of safety of travel.

16. To use materials for the several distinctive parts of the various classes of structures to be built, of the kind and quality preliminarily specified.

17. To buy and pay for all lands to be occupied by the main, secondary, and tertiary canals, and other works forming a part of the system.

18. To pay for, as a servitude, the right of way for smaller ditches of distribution.

19. To pay all indemnities for temporary occupation or deterioration in value of lands, or for the stopping of manufactories pending

construction of any work, and all damages whatsoever which should occur in consequence of such cessations or the execution of works.

20. To maintain, at all times, the principal canal, with its diversions and dependencies, in a good and efficient state of repair and order.

21. To do all that can be done to assure, during the irrigating periods, the full supply of water contracted to be delivered periodically to the irrigators.

22. To do all that can be done to regularly deliver at all times the quantity of water engaged for public or private use, for power, machinery, and industrial purposes.

23. To mark out the boundary of the districts and sub-districts of irrigation, and make complete maps of the same.

24. To survey, stake out, and prepare complete plans of all canals and ditches.

25. To plan, describe, and specify in detail all works entering into the system, before they are undertaken.

26. To pay taxes on lands occupied by all its canals, structures, and other works.

27. To pay taxes on buildings, sheds, and store-houses.

28. To pay taxes on its canals and ditches.

29. To guarantee to deliver, on demand, at times of lowest supply, the full amount of water subscribed for by a certain number of subscribers who subscribe first.

30. To suffer a deduction of rents in case of non-delivery of waters, except as per condition No. 24, following.

31. To suffer roads, railroads, etc., approved by the administration in future, to be built across its works.

32. To employ such agents and guards for the police of the canal, for the supervision of its working, as can be sworn as rural police officers.

33. To bear all expenses of preliminary examinations, surveys, plans, etc., all expenses of construction, etc., superintendence, government examination and engineering, supervision, and examinations for acceptance on completion.

34. To have the headquarters at Valence, there to have a resident agent authorized to receive all government communications and generally transact the business of the company.

35. To deposit within eight days after final organization of the company, in the consignment fund of the treasury of state, under the title of a bond, the sum of 75,000 francs ($15,000,) to be held until the works have progressed to the expenditure of 300,000 francs, ($60,000) as reported by the government engineers, etc.

CONDITIONS OF THE CONCESSION.

The grantee has certain privileges, under conditions as follows:

1. That it (the company) always leaves in the water-courses whence it derives its supply, at lowest stage, a flow below its dams of at least half a cubic metre (17.5 cubic feet) per second.

2. That the individual distributing headgates, drainage ditches, and other such works shall belong to the irrigator in each case, and be built by him or at his expense.

3. That the consumers can compel the company to construct any certain distributing system when they have subscribed for enough water to be delivered by it to guarantee six per cent interest on its estimated cost.

4. That all plans for the main works be approved by the central administration before construction.

5. That all plans for distribution works be approved by the prefect of the department before construction.

6. That plans for all works shall first be approved by the chief of the government civil engineering bureau.

7. That all changes of plans shall be approved by competent authority before the work is executed.

8. That the society shall execute the works under the superintendence of its own agents, but under the supervision and inspection of those of the government.

9. That all works, during the term of the concession, be subject to inspection annually, and oftener if deemed necessary in cases of accident or complaint, by the government engineers.

10. That in all that concerns supply, maintenance, and repairs, either ordinary or extraordinary, upon the failure of the company promptly to act, the administration, through the engineers, may carry out the necessary measures or works at the expense of the company.

11. That the main works will be provisionally received, upon the favorable report of a commission of inspection appointed by the administration, each as completed.

12. That the final reception by the central administration will not take place until one year thereafter.

13. That the report of the commission of inspection be in each case accompanied by full and final plans and reports of the work done, prepared at the company's expense.

14. That two copies in full of these plans, reports, etc., be furnished, one for the department offices, and one for the central administrative offices, at the company's expense.

15. That the same operation shall be gone through with after com-

pletion of the secondary systems of works, but that in these cases the reception be made by the controlling engineers and approved by the prefect of the department.

16. That if within two years after the date of the concession the company has not commenced the main works, it forfeit all rights under the agreement.

17. That if within the term of five years the company has not completed the main works and fulfilled other requirements specified, it forfeit all rights and properties, which are to be disposed of as the government may direct.

18. That, in the event of forfeiture, the company is to receive from the party into whose hands the property goes, a sum to be adjudicated by referees.

19. That, if after two trials at settlement, as to amounts to be paid, there be no agreed result, the company forfeits all, summarily.

20. That forfeiture cannot be enforced if great unforeseen circumstances intervene to prevent the completion of the obligations.

21. That the administration shall determine the duration and time of the irrigation period each year.

22. That irrigation necessities are to be preferred to those of manufacturing.

23. That subscribers may, by payment of a sum to the society, which, at six per cent, will represent the capitalized value of their water rents, thereafter be freed from payment of such rents.

24. That no reduction can be demanded on water rents, should a scarcity of supply result from accidents not to be guarded against by the company.

25. That subscribers are bound to irrigate land at the rate of one hectare (2.47 acres), or less, to the litre (0.03 cubic foot) per second of water subscribed for, and not to divert the water for any other purpose than as agreed upon by the subscription. Nor can any subscription be for a less amount, for irrigation, than one litre per second.

26. That the consumers of the water in the sub-district supplied by each secondary canal may form a syndicate association, under the terms of the law for such organizations, and take out of the hands of the company the works of that sub-district, by paying annually, in bulk, to the company a sum equal to six per cent on the cost of the works, or a sum equal to the water rents subscribed in the district, according to the water demanded.

27. That the company may transfer the works in any sub-district to a syndicate of the consumers for any agreed upon amount; but must thereafter deliver all water subscribed for in the sub-district.

28. That the grant or concession be for a period of ninety-nine years, commencing from the date of the provisional acceptance of the main works.

29. That at the expiration of the time of concession, the company have no more right to the works, but the whole property be turned over to the state in good condition.

30. That, to insure this last condition, the works are to be inspected and put in proper condition, under the direction of government engineers and at the expense of the company, within the two years preceding the expiration of the term of concession.

PRIVILEGES GRANTED TO THE COMPANY.

On the foregoing conditions the company has the privilege:

1. Of taking seven cubic metres (245 cubic feet) of water per second from the Bourne River.

2. Of making up this volume at low stages, by taking two cubic metres (70 cubic feet) from the rivers Lyonne and Cholet. (See condition No. 1.)

3. Of supplying a certain district of 22,000 hectares (54,340 acres) in area, of which 10,500 hectares (25,935 acres) are irrigable, with water for all purposes—irrigation, manufacturing, industrial use, domestic, and municipal purposes.

4. Of doing work and using material of a better class than preliminarily specified, according to the judgment of the government engineer.

5. Of showing the administration at any time why plans of construction should be changed, and asking for changes.

6. Of representing to the administration at any time, conditions or facts which has rendered it impossible to fulfill its engagements.

7. Of shutting the water off from the canals, for purposes of repairs and clearances, for one month each year, at a time to be fixed by the prefect of the department, and not in the irrigating season.

BENEFITS TO THE GRANTEE COMPANY.

And the company is the recipient of benefits as follows:

1. The authority to collect water rents, for the term of ninety-nine years, as follows:

For irrigation—From all who subscribe before the water is put in the main canal, for a fixed amount of water annually, at the rate of 50 francs per litre ($269 per cubic foot) of discharge per second during irrigation.

From all those who subscribe after the water is put in the main canal, at the rate of 60 francs per litre ($323 per cubic foot) of flow, etc.

From the first subscribers above named, for an additional amount, engaged after the water is put in the main canal, equal to that at first subscribed for, at the same rate of 50 francs per litre ($269 per cubic foot) of flow, etc.

For all subscribed for by them over this double of the first subscription, at the rate of 60 francs ($12), etc.

For domestic, municipal, garden watering, ornamental, and other similar purposes—For a continual supply, at rates stipulated.

2. The authority to sell motive power, during the term of the concession, to individuals who want to utilize it for factories, at an annual rental of 200 francs ($40), per one horse-power; a single horse-power being represented by a volume of 100 litres (3.5 cubic feet) of water per second, having one metre (3.28 feet) fall.

3. The authority to collect, under the executive power of the prefect of the department, and in the same manner taxes are collected, all rents for irrigation waters subscribed for, during the last three months of the year, in advance.

4. The authority to collect, in the manner spoken of above, all rents for water for domestic, municipal, and other similar purposes, and for motive power, at the commencement of the year, in advance.

And finally, the government, through the minister of public works, after the company has shown a subscription for water to the amount of 3,000 litres (106. cubic feet) per second, or more, engages to pay the company a subsidy of 2,900,000 francs ($580,000), as follows:

Ten per cent on final completion of all works.

Two thirds of balance on works done or expenses incurred on main canals and works, in installments amounting to one third of actual costs, as reported by the government engineers.

The other third, in the same way, on works of the secondary and distributing systems, etc.

SECTION II.

REGULATION OF WORKS.

GOVERNMENT IMPROVEMENT OF NAVIGABLE RIVERS.[*]

The rivers of France generally have high rates of slope and rapid currents, where works of irrigation and water power are constructed.

---
[*] See, Debauve, and articles referred to by him in "*Annales des Ponts et Chaussées;*" also, Malapert.

The channels are through heavy formations, as compared to the alluvions of Californian valleys, and the beds are almost always gravelly, and not infrequently rocky.

Such streams may in their upper courses pass through alluvial irrigable valleys, and then again meander through ravines and rolling lands.

It is due to these characteristics of the hydrographical system that water power early came into very extended use in France, and, following the development of trade thus caused, that the demand arose for making the streams themselves navigable.

Thus, the system of canalizing rivers by means of dams, in series, at intervals along their course, making nearly slack water navigation between each two, naturally came into being, and has resulted in a high degree of skill and perfection of practice in the general disposition of such works and arrangement and construction of their parts.

The French masonry and iron frame movable dams of several distinct types and patterns, are models of construction in this line for engineers of other countries, where similar conditions obtain and like purposes are to be subserved.

The government civil engineers have charge of such rivers throughout their valley course, and it is the endeavor to bring all works into harmony with a system best calculated for the public utility of the streams and the safety and well-being of the interests along their banks.

EXTENT AND FIELD OF THE HYDRAULIC SERVICE.*

The hydraulic service of the public works bureau comprehends the supervision of river bank and channel works relating to the creation of power for manufactures, diversion of water for industrial or other similar uses, for irrigation and *colmatage*,§ the cleansing or dredging of channels, improvement of channels, construction of embankments and other defenses against floods, draining of marshes, sanitary improvement of moist lands, and agricultural drainage.

By the very nature of the objects contemplated, the service is divided into two sections—one dealing with those cases where the water is an auxiliary in the purpose held in view, the other with those cases where it is an enemy to be encountered in effecting the desired end.

Works connected with manufacturing, industrial, and other uses, irrigation, and colmatage, fall in the first section, while all others mentioned above naturally rank in the second.

---

* See, De l'Assy, Malapert, p. 417 and elsewhere.

§ *Colmatage* is the French word for warping, silting-up, or enrichment of lands, by leading muddy waters upon and causing the silt to be deposited on them. It is extensively practiced in many quarters of France, Switzerland, and Italy.

REGULATION OF WORKS ON RIVERS.

THE PRINCIPLES OF COÖPERATION AND COMPULSION.*

When the water is an auxiliary, enterprises are frequently carried out by individuals, as in manufactories, etc., and always by voluntary action. While in irrigation and colmatage enterprise, the initial movements are frequently on the part of collective interests, but always voluntary, so far as each individual at interest is concerned.

When, on the contrary, the water is an enemy, as in the improvement of channels, sanitary drainage, works of defense against floods, there is always an indissoluble common interest at stake, so that the movement must be for the collective benefit of all land within some certain district, primarily, and for that of the public generally as well, or else not for the benefit of any. In these cases the enterprise must *necessarily* be on the part of the collective interest of all concerned, and the law submits the minority to the will of the majority of interested land holders in the district. "It cannot be allowed," says De Passy, "that enterprises so essential to agricultural development be defeated by the resistance or indifference of an ignorant and capricious minority."

Moreover, if it is recognized that the enemy to be fought inflicts injury on the public interests, the administration has the right to interfere and render obligatory the common action of all interested parties in the district, in spite of the opposition even of a majority.

REGULATION OF THE CONSTRUCTION OF DAMS.§

Whenever possible, the holding up or diversion of water for a manufactory, an irrigation canal, an industrial establishment, or other use (requiring the construction of a dam in the river and acquirement of elevation in the water plane to give a head for power or for flow out from the channel) is effected by a work which serves at the same time to hold back water for the promotion of navigation.

The height of such a dam is limited by the elevation of the plane of safety to the lands which might be flooded by backwater were it carried too high, and, at the same time, it is governed by the requirements of navigation for a certain depth of water in the reach above.

The cost of such works, in so far as they relate exclusively to navigation, is borne by the state; the grantee of the water privilege, for whatever purpose the use may be, exclusively bearing the cost of his sluices and gates.

When, however, dams are designed and constructed for the common benefit of navigation and some water privilege establishment,

---
* See, De Passy, pp. 7 to 11.
§ See, De Passy, pp. 299 to 324.

they are paid for and maintained at the joint expense of the state and the water grantee, in proportion to their respective interests, unless special ancient agreements determine the distribution.

The distribution of expenses for construction, as well as for maintenance of works built conjointly by the state for navigation and water grantees for their purposes, is made before the works are executed, in every case by the central general administrative authority—the whole council of state in general assembly—and is promulgated in an administrative decree.

The grantee's part of the cost is fixed at a sum to be paid annually, and not in a sum paid at once. Thus the coöperation of the grantee with the government results in his paying an annuity for his benefits from the construction, and not in his paying at once, in part for the work itself, and thus acquiring a right of property in it; for works of this character, forming a part of the essential system for navigation, must remain always public property.

The determinations of the council of state in these matters are based on the reports and estimates of the government civil engineers, and are also shaped in accordance with equities arising from the peculiar circumstances of each case, taking for comparison, if need be, the results of other similar works carried out under parallel circumstances.

Whatever is paid by the water-privilege grantee, goes into the coöperation fund for public works, under the control of the minister of public works.

Such works are built and repaired and wholly cared for by the administration, and, as far as necessary, under the advice or direction of the government civil engineers.

In cases where a new dam, not necessary for navigation, is to be established for the benefit of a water-privilege grantee, he is obliged to provide in his plans and construct at his expense, a proper lock for the passage of boats.

Should the administration recognize in the work a benefit to the river navigation, the government may contribute to the cost of the lock.

Plans for works constructed by grantees alone, are always subject to revision by the government civil engineers, and the carrying out of such works is subject to their inspection and approval or condemnation.

REGULATION OF THE CONSTRUCTION OF HEADWORKS.*

Works designed for taking water for any purpose of a holder of a water privilege are always constructed and maintained at his expense, and when in close connection with a dam for navigation purposes, are carried out by the administration, or under the immediate supervision and superintendence of the government civil engineers, or, if not connected with a navigation dam, they are subject to supervision only, the plans having been approved.

As waters for manufacturing, irrigation, and other grantee purposes (except in the case of supply to municipalities for domestic purposes), can only be drawn from the excess of supply over demand for navigation purposes, the determining and gauging of the quantity allowed, so that at times of scarcity proper equity may be observed in apportioning the available surplus, becomes a matter of extreme importance.

The forms and dimensions of the sluice ways, or gate openings, the elevations of the sills, with respect to that of the dam's crest, and the legal low water plane of the river, always form the subject of a special clause in the decree authorizing the establishment of the works, and hence any modification in the plan of a dam or headwork intended to divert water, can not be made until duly authorized by government.

If the quantity of water to be taken in any instance amounts to a considerable volume per second, as is commonly the case in works intended for irrigation, it becomes necessary, in providing for the regulation of the discharge, not only to determine and fix the size and form of the head-gates, but also the form of section and gradient of the canal or other water-way leading therefrom for a certain distance varying with its size.

"For, in all cases, it is to be remembered that the sluice for taking water is the sluice for guarding it."

If the quantity to be taken is small, in the case of irrigation, it is deemed sufficient to provide for taking it through a culvert or pipe of determined area and under a fixed head.

In cases where water is delivered in rather small quantities for distribution by sale, it is parted out into a "sump," and then more accurately measured over a gauge weir, of which the crest is arranged so as to preserve a fixed head of water producing the requisite discharge.

---
* See, De Passy, pp. 306 to 316; also, Dumont.

SECTION III.

OPERATION AND MAINTENANCE.

GENERAL MAINTENANCE OF WORKS.

Concerning the subjects of this heading, very much has necessarily been said under those which precede, nevertheless, it will be well to call attention to some leading points already mentioned, in connection with matters not yet spoken of.

The care of all navigable streams in France is committed to the administration; all public works pertaining to the stream as a navigable channel, or as a drainage way of the country, are in care of the officers of the hydraulic service, and their assistants and subalterns. These officers are, as a general thing, civil engineers, holding commissions as such, and are under the government public works bureau.

The maintenance of all private works bordering upon, or in such streams, and calculated to affect them as navigable channels, or as natural drainage ways, is subject to conditions imposed in terms of the grants of privilege, and subject to the general and particular regulations of the administration, as executed by the officers of the hydraulic service.

Works of navigation, are, of course, maintained and operated solely by the government, the tolls on navigation, which are very low indeed, defraying these expenses.

Works built on joint account of state and private enterprise, are maintained and operated under government direction, at joint cost according to prefixed agreements, or as may be equitable under the circumstances, or, again, as may be customary from ancient times.

Works solely for the benefit of private interests are maintained under administrative supervision, at the expense of the owners, and if the work is not properly and promptly done, the administration, if public or communal interests are threatened from negligence or faulty construction, may carry it out at the expense of the owner or responsible party.

Besides the special and local operations of maintenance applicable to works on the streams, there is the care of and cleansing or dredging of the channels themselves, and the police of their banks.

CLEANSING OR DREDGING OF CHANNELS.[*]

The necessity for cleansing the channels of water-courses arises largely from the effects of natural causes, such as abrasion of stream

---

[*] See, De Passy, pp. 323 to 328; also, Dumont.

banks and denudation of lands drained; but artificial causes, such as deposits from boats, and from the shores by the inhabitants, by towns, and industrial establishments of all kinds, contribute largely to the results.

The dams built in the channels for the promotion of slack water navigation, or for the creation of power heads for manufactories, or for whatever purpose, prevent the scouring of the bed, and serve to cause deposits of sediment and filth that otherwise would be carried away by the currents.

Upon navigable and raftable channels, of which the bed and banks are public property, the clearances are made chiefly at the expense of the State.

When the dams on such streams are used to create water heads for power purposes, as well as for navigation, the holders of the water rights are called upon to pay part of the expense.

When the administration believes that the cleaning work is necessary only in the interest of navigation or raftage, its cost is borne solely by the government.

When the clearings are necessary solely in the interest of public health, and are made necessary by the deposit of filth in the channel, from towns, residences, and establishments on the banks, the expenses are charged for the most part to the riparian owners and the towns, and in a small degree to the state and the manufacturers whose dams increase or favor the deposits.

Such cleansings are ordered by the superior administration, which determines the basis of the work and the distribution of expenses, on the reports of the engineers and local administrative officers.

Upon non-navigable and non-floatable water-courses which have not been declared to be dependencies on the public domain under article 538, civil code, and which have not been improved in the interest of navigation, the expense of cleaning and caring for the channel generally, is borne principally by the riparian land owners, as will be seen in the next chapter.

POLICE OF STREAMS.[*]

Works erected and acts committed in the channels or on the banks of non-navigable or non-raftable water-courses, when they present no obstruction to free flood-flow, as they only give rise to questions between private interests or individuals, are subject only to regulation by the law as administered by the courts.

In these cases it is necessary only for the administration to examine

---
[*] See, De Passy, pp. 326 to 334, and elsewhere; also, Dumont, De Buffon, and Malapert.

the project with the view to determining whether or not the stream channel or the public interests are likely to suffer, or the flood plane likely to be affected by its results.

Works located upon navigable or raftable streams when not duly authorized by the administration, constitute infringements of the laws of the commission of public ways, and are subject to repression by the council of prefecture.

The legislation in the matter of police of public water-courses and canals is found in the judgment of the council of state of the king, dated twenty-fourth June, 1777, confirming and completing former rules, notably those of forests and waters, dated August, 1669.

The various articles of the judgment of 1777, specify the penalty attached to each kind of offense enumerated.

Besides this old general law, there still exist in force a number of ancient special enactments applicable to the principal rivers and to certain navigation canals, emanating from the king in council of state, from the governors of provinces, and from other authorities who under the ancient régime exercised the ruling power.

Other ruling enactments on this subject bear dates subsequent to the revolution, but none of them are of recent origin except that of twenty-third March, 1842, although there are many decisions under these laws, that interpret and modify their application.

The penalties fixed in the old laws were very severe in proportion to the offenses to which they were attached, and the councils of prefecture, in the administration of the laws, had no alternative but to apply them in full vigor, for the mitigation or repression of such penalties could only be authorized in each particular case on an appeal to the chief executive power of the council of state.

The law of 1842 gave to the councils of prefecture the authority to gauge the penalties to the offense in each case according to circumstances, between 16 francs ($3 20) as a minimum and 300 francs ($60) as a maximum for the generality of ordinary offenses.

Works having a direct effect to the detriment of public interests may be summarily removed on the order of the prefect, and formerly unauthorized works on public water-courses, whether injurious or not, could be similarly disposed of without delay. But now in cases where no injury is done or immediately threatened, a delay for a reasonable time is granted to give the owner of the works time to appeal to the superior administration for a proper authorization for his enterprise.

The administration of all laws governing the police care of public streams in the interest of the public, whether protecting navigation or other particular interest, is left to the prefects of departments. But a large class of cases, where the laws have to be interpreted, and where private interests are affected, find jurisdiction before the courts.

WATER PRIVILEGE RENTS.*

Every concession of a water privilege on streams of the public domain is subject to the charge of an annual rental which goes into the general treasury of the state for the benefit of the public works.§

In the case of water heads for manufactories, the rent is based upon the purchasable value of the gross power conceded, independent of any special advantage which the grantee may get from it, and of the kind of employment to which it may be devoted.

The rate of rents for manufactory water-powers is a sum per annum equivalent to one two-hundredths of the purchasable value of the motive power measured in horse-power.

The purchasable value of the horse-power is determined by precedents on the stream in question, and on other similar streams where water is used for like purposes.

Water privilege rents for irrigation works are rated upon the basis of the increase in yield due to irrigation, and are fixed at a sum annually paid, equivalent to one tenth of the increase in value of produce on the land irrigated over its produce before irrigation.

Industrial purposes include all the purposes of manufacturing, except that of creating motive power by means of water wheels; thus water for making steam, for condensing steam, for the use of paper mills, sugar refineries, tanneries, bleaching works, cloth printing works, etc., is ranked as used for industrial purposes.

Whether taken by means of pumping machinery or not, if the volume of water in any instance drawn directly from a public stream for an industrial use is sufficiently large, in proportion to the supply at any season, to sensibly affect, or, in the opinion of the engineers of the administration, injure the normal régime of the stream, the water privilege is ranked with those for water-power purposes.

For all concessions of water for industrial purposes, the basis of annual rental is a fixed sum which is adjudged for each particular case, the minimum being one franc and an additional ten centimes per cubic metre or fraction thereof of water taken per day.

Water heads for municipal domestic purposes are governed by the same rules as those for industrial purposes.†

---
* See, De Passy, pp. 306–307, 314–316, and elsewhere; also, De Buffon.
§ Financial Laws. June 16, 1840; July 14, 1856.
† Decrees March 25, 1872; April 13, 1861.

When the object of the works is simply the supply of domestic requirements, without revenue being derived by the sale of the water to consumers, the rent is fixed at the nominal sum of one franc (20 cents) per year; the object being merely to assert and maintain the right of the state to regulate and control such matters.

When the intent of the grantee, whether a town or a company, is to sell the water to consumers and derive a revenue from it, the case is ranked as an industrial use, and in addition to the fixed amount of one franc, a charge of ten centimes per cubic metre (35 cubic feet) of water drawn daily, is imposed.

The amounts of all annual rentals are based on the reports of the government engineers as to volumes diverted and according to gaugings and records, and when a gauging is made and a record is kept, the grantee is obliged to assent to its correctness or at the time show it to be erroneous.*

Back rents for water can be collected for five years, but recovery for a longer period of time is debarred by a statute of limitations.

All questions as to rates for rents are considered by the ministers of public works, and of finance conjointly.

Without meaning in any way to limit the duration of water concessions, the rents are revised every thirty years, for, although revokable at any time, water-right concessions on public streams are given for an indefinite time, and in most cases practically for ever. Any other system would be opposed to the development of industrial prosperity.

Water privilege heads held in private control previous to the edict of 1566 declaring the inalienability of the public domain, are free from the charge of rents, as are also those whose holders have titles derived by purchase from the government.

AUTHORITIES FOR CHAPTER III.

In the preparation of this chapter I have consulted the following named authorities:

Dumont.—[Work cited as an authority for Chapter II.] See Book II, Chapters I, II, and III.

De Passy.—[Work cited as an authority for Chapter II.] See pp. 7-11; supplement, pp. 297-334.

Malapert.—[Work cited as an authority for Chapter II.] See the headings, "The Actual Republic," and "Engineers."

De Buffon.—[Work cited as an authority for Chapter II.] See, generally, Vol. II, Part II.

Reclus.—[Work cited as an authority for Chapter II.] See, generally, descriptions of France.

Debauve.—Vol. XIX. [Work cited as an authority for Chapter II.] See, generally, description of river works and systems.

Les Annales des Ponts et Chaussées.—A semi-official publication of the French Government Corps of Civil Engineers; comprising volumes of Technical or Engineering matter, and others of Laws and Decrees relating to Public Works and the Engineering Service, generally. See late volumes, and, particularly, Vol. CXXVI, pp. 451 et seq.

---

* Decree of the Minister of Finance, May 15, 1863.

CHAPTER IV.—FRANCE[3];

WATER-RIGHTS ON, AND THE ADMINISTRATION OF NON-NAVIGABLE STREAMS.

SECTION I.—*Rights to the Use of Water.*
    Water-rights previous to the time of the Code Napoleon.
    Riparian Water-rights under the Code.
    Nature of the Riparian right, and tendency of interpretations.
    The right of Irrigation—absorption of water, etc.

SECTION II.—*Supervision of Construction of Works.*
    Decentralization of the Administration.
    Powers and Duties of Local Administrations.
    Applications for sanctions to construct Works.
    Obligations and Conditions attached to Permits.
    Construction and Regulation of Dams and Headworks.

SECTION III.—*Regulation and Operation—Works and Waters.*
    Necessity for Regulations and Administration.
    Administrative Authority to make Regulations.
    Principles adhered to in making Regulations.
    General Rules as to Division of Water Supply.
    Regulations of Irrigation.
    Division of Waters between Claimants.
    Regulations for Streams.
    Police and Cleansing of Water-courses.

SECTION I.

RIGHTS TO THE USE OF WATER.

WATER-RIGHTS PREVIOUS TO THE TIME OF THE CODE NAPOLEON.*

As we have seen, streams not navigable nor floatable—those upon which tolls could not be collected for navigation or rafting facilities, or heavy rents derived from ferrying franchises—having been claimed and controlled, together with all other water-courses, by the feudal counts during the early centuries of modern ages, were also included in the property-right claim of the kings, and originally contended for by them against the counts; but in the course of time the struggle was made only for the control of the larger water-courses, from which

---

* See, particularly, Dalloz, Vol. XIX, pp. 312-319, and Dumont; also, De Passy, and Malapert.

revenues could be derived, and those of the smaller class were left to the owners of the lands adjoining them.

Matters appear to have rested in this way for a long time: the exclusive right to water, for milling and irrigating purposes, from streams too small to be regarded by the kings as of public importance, according to the standard of the times, being accorded to the owners of the bank-lands, apparently upon the ground that they owned the beds and waters as well as the banks.

In later years, when it was found necessary in the public interest, and to rid the courts of a vast volume of litigation, for the government to supervise the placing and maintenance of structures in the channels and the diversion of waters, it appears to have become recognized that the waters were in reality a common property, and that the bank proprietors had only a right to use them and not a right of ownership in them.

Still there was the open question, to whom were the waters a common property: the riparian proprietors claiming to be the owners in common of the waters of each stream, and submitting to the control of the streams by the government only as it was based upon the general police authority of the nation; while the government asserted its right to control, not only because of its general police powers, but because of the fact that the waters were really the common property of the whole people and not of the riparian proprietors alone, and, that public interests were to be promoted as well as other private interests guarded by it, and that, hence, its mission was one to promote public utility as well as to repress or prevent abuse of private privileges in the protection of other privileges.

CONFLICTING INTERESTS ON THE STREAMS.

The continued and growing abuse of the riparian water-right privilege brought about an increased necessity for upholding this latter view, so that from having been a governmental administrative measure it became a popular sentiment, and owners of lands not riparian to the streams asserted a right to waters for their irrigations, on the ground that such waters were a common property of all the people; and asserting that the riparian owner's privilege of using them was not an exclusive privilege, but that upon a grant or permit from government, any land owner could divert them for use on his lands.

In this view of the case by far the greater number of land proprietors were interested, so that the governmental policy of control was strongly upheld.

But now manufacturing interests, which were wide-spread and

becoming powerful, took alarm. The owners of the hundreds of mills and manufactories depending on water supply for power and other purposes, scattered along the streams all over France, and holding rights, many of them dating back in the times of the counts, and all valuing the riparian right as a protection to their water supply, were arrayed against the advancing theory—of the waters belonging to all the people and due to all the people for use.

The government from time to time brought to face the question in deciding points at issue, continued to uphold the theory of the waters of these small streams being a common property of all the people, and framed its own measures accordingly, but no step was taken to accord land owners other than riparian proprietors any right to use them.

RIPARIAN WATER-RIGHTS UNDER THE CODE.[*]

The case appears to have stood in this way when the Code Napoleon was promulgated in 1804.

This code contained provisions (articles 713, 714) which in course of time were recognized as placing the ownership of the waters of the smaller class of streams in the nation, but declared the use of things of this class to be common to all.

Left with this provision only, the waters of these streams would have been thrown open to use by all the people; "the laws of police regulating the manner of enjoying them," as the code said.

But article 644, under the head of servitudes, seemed to place a special servitude (right to use) on these waters for the benefit of riparian estates. It reads as follows:

"He whose property borders on a running water, other than that which is declared a dependency on the public domain by article 538, under the title 'Of the Distinction of Property', may employ it in its passage for the watering of his property."

"He whose property is intersected by such water is at liberty to make use of it within the space through which it runs, but on condition of restoring it at the boundaries of his field to its ordinary course."

The provisions of this code have given rise to many questions, or rather to the old questions in new form, accompanied by an infinite number and variety of side issues. The old question as to whether or not the riparian water-right privilege was an *exclusive* right, was left open with additional complications.

The government had its hands strengthened in its policy of control and regulation, and the fundamental principle contended for by the owners of lands not riparian to the streams, as well as by the government, was recognized.

---

[*] See, particularly, Dumont, pp. 171-208, and De Passy; also, Dalloz, Vol. XIX, pp. 379-390.

But riparian proprietors claiming and being, in some cases and under some circumstances, allowed ownership of the beds of the streams, still claimed ownership of the waters by virtue of article 552, which reads: "Property in the soil imports property above and beneath."

And a stand was thus made by riparian interests, on the point that as the waters of the streams rested on their lands, they belonged to them, and, hence, articles 713 and 714, about "things which belong to no one," had no application to them.

Article 645 provided expressly for the settling of disputes which should arise under the preceding article at least, in the following language:

"If a dispute arise between proprietors to whom such waters may be useful, the courts, in pronouncing their judgment, must reconcile the interest of agriculture with the respect due to property; and in all cases, particular and local regulations on the course and use of waters must be observed."

Under this article, all questions as to rights to use waters from nonnavigable and non-raftable streams have been carried before the courts, and these have not directly recognized the claims of the back land owners, leaving the riparian proprietors in possession of the field.

The central administrative authorities were appealed to to exercise their authority in behalf of the land interest which had sustained its authority and theory of public ownership and government control of the waters.

But the administrative authorities have consistently replied to these appeals that, under article 645, they had no jurisdiction in this class of cases: that the courts were the only resort of those claiming water in this class of streams, in which to make good a claim.

THE RIPARIAN WATER-RIGHT AND THE RIGHT OF WAY.*

Another point which for a long time was in favor of the riparian proprietors, was the fact that there existed no law under which a back land owner could get the right to conduct water over the property of those between him and the stream, even though he had the right to it, and no law under which he could get the right to abut a dam against banks belonging to others, even if he could get the right of way by amicable purchase, and the water also, and, furthermore, the administration could not grant such privileges.

Companies or syndicates contemplating extended irrigation enter-

---

*See, particularly, Dumont, pp. 225–256, 259, 280, De l'asay, Dalloz.

prises were granted water privileges and the right to construct works by decrees of the central administration, and their works being declared of public utility, they were authorized to condemn by process of law the right of way for their main canal. But no single land owner, and no enterprise not declared to be of public importance, could get right of way, except by private negotiation.

In 1845 a law was passed giving land owners generally the power to secure rights of way to conduct waters to which they had a right of use, as a servitude, over lands not their own.

This was ostensibly in the interest of riparian proprietors who had to take water out of the streams above their own lands to get it high enough to conduct on to them. But it was also a step in the direction of the theory of the back land owners.

In 1847 a law was passed giving the owner of one bank a right to abut his dam against the bank owned by his opposite neighbor, under certain regulations and administrative sanction, etc.

This also was a step towards breaking down the exclusiveness of the riparian right to the stream.

Until within the past few years a riparian proprietor, upon the basis of his claim of ownership of the banks and bed of a stream, so far controlled the channel, as against other private individuals, themselves also riparian proprietors, as to deny the right to construct a dam below in such manner as to back the water up into the channel opposite his land, even though there was no apparent material injury to him caused thereby.

But now the court of cassation, at the head of the judiciary of the country, and the council of state, at the head of the advisory department of the executive branch of government, have each decided that "the fall or slope of a channel is not the property of the land proprietors, and that it enters into the class of things which by the terms of article 714, Code Napoleon, do not belong to anybody, of which the use is common to all, and of which the enjoyment is regulated by the police laws;" and the administration grants a proprietor the right to back water into the channel in front of lands above him, by means of his dam, so long as he does not injure or endanger the lands in any way, take away from the efficiency of other works above, or endanger the public interest.

Here again was a step towards the abolition of the exclusive and complete riparian control of the stream, and a movement towards a declaration of public ownership of the channels themselves.

And thus the matter stands. The riparian proprietors still monopolize the right to use the waters from streams of this class; indeed, in this respect they have an exclusive and complete right as against all comers, except "public utility," "public health," and "national welfare."

"To exercise the right of irrigation, it is necessary to be a riparian proprietor. If, then, a water-course comes to change its bed, the ancient proprietors, who are no longer on the new bed, no longer preserve upon it the right of taking water for irrigation, nor, consequently, of making constructions destined to conduct the waters upon their properties."—[Dalloz, Vol. 40, word "Servitudes."

The administration, representing the whole people and the nation, by virtue of its police powers and its guardianship of public property and public weal, exercises a control over the streams, a regulation of all works placed in the streams, and a surveillance of all use made of the waters.

NATURE OF THE RIPARIAN RIGHT, AND TENDENCY OF INTERPRETATIONS.[*]

The nature of this riparian right to water on non-navigable streams in France may be a little difficult to comprehend.

It is so far a right to have the water left in the channels that the administration, on the ground of "police regulations," "sanitary provisions," or "public utility," may refuse to sanction the construction of a work for diversion, which has not proper provision in the way of sluice-gates to let water enough go on down stream for domestic purposes of all bank owners below, at the driest times; nor will it sanction the construction of a dam, when it appears that owners below will be deprived of water by its effect, although the projector be a riparian proprietor and has a right to water under the code.

And yet, there is no element of the principle of prior appropriation—first in time first in use—about this right.

The code dedicates these waters to the use of him "whose property borders on," or, "whose estate is intersected by such waters."

It is only in the regulation of affairs by the courts and the administration that any recognition of priority of right is found, and even then, in the supervision of the use, the principle is not closely adhered to.

The code merely gives every riparian owner a privilege of using the water. There was no recognition of old and established rights in this connection, although many such existed; nor any rule laid down except that "in all cases particular and local regulations on the course and use of waters must be observed," and that "the interest of agriculture" must be reconciled "with the respect due to property."

---

[*] See, particularly, De Passy, pp. 23, 24, Dumont, pp. 171–208; also, Dalloz, Vol. XIX, pp. 379–390, and the *Annales des Ponts Chaussées*, vols. Laws and Decrees (recent).

And yet this riparian privilege is so far a right to take water out of the stream, that, though fully used, the courts can recognize a right for a new water privilege, and the administration may sanction the works necessary for availing of it, and, in the course of the division of waters, the new work will get its share.

This rule, however, would not be carried so far, presumably, as to deprive any user of water, of all he actually required to accomplish his purpose, but it would force him to economize in his use.

No matter how old a privilege may be, the administration in the public interest has always the right to turn sufficient water past the dam to satisfy the personal wants of proprietors below, and thus guard against unsanitary results; and it can even compel the construction of a sluice-way in the dam, to be used for this purpose.

THE RIGHT OF IRRIGATION—ABSORPTION OF WATER, ETC.[*]

For many years after the promulgation of the code it was held that the obligation imposed upon the riparian proprietor of "restoring it (the water) at the boundaries of his field to its ordinary course," after use, as set forth in the second paragraph of article 644, applied as a condition to all use of water allowed by the article, and, hence, there could be no material loss by absorption in irrigation.

The irrigations in France at that time were very generally those of meadow lands situated closely along the stream borders, and a very large proportion of the waters run on to them flowed off again.

The court of cassation (Supreme Court of France), in 1844, August 21, rendered a decision on this point as follows:

"Running water is regarded by the law as a common property. Riparian proprietors on a water-course naturally have equal rights to the use of the water, although they cannot exercise this right simultaneously.

"If on account of the advantage of its topographical position the proprietor of higher land on a stream, exercises his right before the proprietors of lower lands, he is not the less obliged by this position after having used the waters, in the interest of agriculture and industry, to return them to their usual bed, in order that the proprietors of lower lands may use them in their turn.

"When the proprietor of the higher land possesses at the same time both banks of the stream his right is more extended; he can then turn the water-course from its bed within the extent of his domain, and take the waters for use where he wills on his estate, being obliged to return them to their ordinary course where it leaves his property.

"This proprietor will not have to return the same quantity of water which he has received, or any certain quantity of water determined, but he must economize and use water in a just measure so that the

---

[*] See, Dumont, De Passy, and Dalloz, as already cited; but particularly, late volumes of Laws and Decrees of the *Annales des Ponts et Chaussées*.

proprietors of lower lands may exercise their rights also."—[Decision—August 21, 1844.

Again in a decision, rendered in 1847, the same court decided that an upper proprietor, no matter how extended his estates on both banks of a stream, had not the right to absorb all the water on his lands, to the detriment of a lower proprietor, and that the lower proprietor had a right to a regulation whereby he would be assured a part of the supply, in accordance with his needs and rights as adjudged by experts. *

THE QUESTION, ONE FOR EQUITABLE ADMINISTRATION.

De Passy, writing in 1878, and a semi-official book for the information of the members of the national hydraulic service, as well as for general sale, says:

"An obligation on the irrigator to return the water when it leaves his lands, to its natural channel, does not result from article 644 of the Code Napoleon. That article comprises two paragraphs, distinct and independent from each other; the first regulates the right of irrigation, which may be exercised by the proprietor of one bank; the second recognizes in the proprietor of both banks more extended rights, such as industrial use, etc.; and it is as a restriction on these last rights, and in the second paragraph only, that is written the obligation to return the water upon its exit from the lands traversed, to the natural channel."—[De Passy, p. 50.

As a matter of fact, the streams are controlled, and the waters apportioned out to those who have claims on them, by administrative regulations. Economy in their use is enforced, according to the experiences of the country; so that the question is kept out of the courts more than it used to be, and the courts recognize the fact that they can make no decision that can settle the point on principle, or even in any particular case for all contingencies that arise. The later decisions are not decisive as to principle; they lean towards the view above quoted from De Passy, speak of "returning the drainage and residue of the waters," only, to their natural channels, uphold ancient customs in the use of waters, but enforce administrative regulations that look towards economizing it, and other measures in the public interest.

---

* Decision—July 8, 1847. See, *Les Annales des Ponts et Chaussées*, Laws and Decrees, 1847.

SECTION II.

SUPERVISION OF CONSTRUCTION OF WORKS.

DECENTRALIZATION OF THE ADMINISTRATION.*

By an imperial decree made in 1852, and interpreted by a number of decrees of the council of state of later dates, a portion of the authority theretofore expressly reserved to the ministers and council of state in matters pertaining to the regulation of water-courses, was delegated to the local departmental administrations. §

This transfer of power constituted what is known as the decentralization of the administration in the hydraulic service.

By it much more responsibility has been thrown upon the engineers, seeing that the scope of their duty has been widened, and some other inquiries being done away with, those which they make must necessarily be more searching, and there being no certain revision of their opinions by a higher central body, their views must be more firmly grounded on good judgment.

The law, however, provides a right of appeal from the decrees of prefects and opinions of the engineers, so that parties being aggrieved at a result may take their case immediately before the minister of public works or even the council of state for revision.

POWERS AND DUTIES OF LOCAL ADMINISTRATIONS.†

The prefects of departments have the power (1) to authorize upon non-navigable and non-raftable streams, the building of all new works necessary for mills, manufactories, dams, headworks for irrigation, etc.; (2) to regulate the existence of such establishments where already constructed without formal permit and regulation; and (3) to modify existing rules concerning such establishments already built.

In these cases the prefects act directly, by simple resolution, without the special intervention of the minister of public works, but upon the opinions and advice of the chief engineers of the departments, and in conformity to the general ministerial regulations and circulars of instruction. §§

They also have the authority to carry out ancient rules and local usages in the matter of the division of waters, from streams of this class, between the various interests employing them.‡

---

* See, De Passy, preface, and elsewhere.
§ *Decrees*—March 25, 1852; April 15, 1861; August 26, 1867; March 18, 1868. *Law*—June 21, 1865.
† See, De Passy, pp. 14, 15, 60–68, 73, and elsewhere.
§§ *Decree*—March 25, 1852.
‡ *Decree*—April 15, 1861.

But in the absence of ancient rules and local usages to serve as a basis for prefectorial regulation of the division of waters between claimants, and especially between antagonistic interests such as manufacturing and irrigation, the prefects have not the authority to act, but such regulations must emanate from the council of state by decree.*

Hence, the prefects can authorize the works necessary for an establishment, but cannot, in apportioning water to it, alter or amend existing regulations concerning the division of waters, so as to affect the interest of others, or the public interest, or change "local usage" in this regard, to the prejudice of third parties, unless there is in existence some "ancient rule" applicable to the case which authorizes the setting aside of such "local usage" by the prefect.

NATURE OF THE POWER HELD BY PREFECTS.‡

The authority of the prefects in the matter of regulating watercourses and waters is confined to the authorization of works, and to the execution and adjustment of details of decrees regulating the distribution of waters, and the application of ancient rules and local usages.

The first power is that of authorization, all the others are in the nature of police powers. Hence, except in the one class of cases mentioned (the authorization of works on non-navigable watercourses), all the regulative measures of prefects are based on police powers, and limited by the ideas of public safety and welfare to be attained by such measures.

The police power is not to be confounded with the power of authorization. The right to take measures in the interest of public health, for instance, has always belonged to the prefects.†

"The nature of police measures consists solely in securing a respect for the public interest, in calling on each person for the execution of his obligations, for the cause of the right and the good of all."—[De Passy, p. 70.

The original declarations of authority, under which this power is to be exercised by prefects, is found in laws of 1790 and 1791, and a resolution of 1799.

The first law charges the administrations with the duty of "seeking and indicating the means of procuring the free course of the waters of streams, with a view of preventing the plains from being submerged by the too great elevation of milldams and of other works

---

* *Decree*—August 26, 1867.
‡ See, particularly, De Passy, p. 15, and elsewhere: Dumont; also, De Buffon.
† *Decree*—March 18, 1868.

established on the rivers, and of directing, in fine, all the waters of their territory towards the one object of general utility, in accordance with the principles of irrigation."

The second law imposes upon the departmental administration the duty and authority to fix the height to which dams may be built in streams, so as "to hold the waters at a height which does not injure any one," or in any way "interfere with the public interest or convenience."

And the third law delegates "to the administrations of departments the power of taking all the necessary steps to prevent waters being turned from their natural courses by works of diversion, simple ditches, or otherwise, without previous authorization; and, also, the power of seeing that dams, embankments, and other works do not exceed the level which will have been fixed for each." *

The duty of prefects in this connection is sufficiently apparent from that which has been said respecting their authority and power, and from what is said under subsequent headings in this chapter.

APPLICATIONS FOR SANCTIONS TO CONSTRUCT WORKS.§

In cases where water is to be taken from a stream without constructing a dam, by a simple cut in the bank, with a headgate, permission to construct the work is not necessary from the administration, for it can only interfere when the flow of the stream is to be checked by a dam,† but the owner of the proposed structure must establish in the courts his right to water, if this be contested, and the construction afterwards comes under the supervision of the administration in carrying out regulations for all diversions and uses on the stream.

But in the interest of the public the administration may cite parties proposing or executing such works to appear in court and prove their right to water, and prove that they will not destroy interests already grown up.

Whenever a work is to be constructed in or on the bank of a nonnavigable stream, which will or may affect its regime as a drainage way of the country, or which may directly affect the common rights or public utility subserved by the stream, sanction of the plans and project must be had from the departmental authorities.

Application must be made to the mayor of the commune, the sub-prefect or the prefect, for the permit, and this application must be

---
* Resolution of the government, 19th Ventose, year 6.
§ See, particularly, De Passy, supplement, and Ch. I; also, Dumont.
† The latest regulations of the administration conflict with this doctrine. See Article 6 of the form of regulations at end of this chapter.

accompanied with a plan of the proposed work, a statement as to its purpose, etc.

The mayor publishes this application by posting it as directed by regulations. He hears and records the substance of all comments or objections, and he transmits the statement of the case to the sub-prefect.

This authority after consideration reports the case to the prefect, who submits the question to the departmental engineer on the special service.

The engineer examines the matter to see that the works are such as will not bring harm to the stream, and in conformity to general regulations. He may prepare other plans to effect the same purpose, and recommend them in place of those contained in the application. These results with his opinion are reported to the prefect, who may order a further investigation of the whole matter by the sub-prefect, or may thereupon act by granting or refusing the application.

To every such permit conditions are attached, binding the grantee to construct the work according to plans or to modifications thereof to be approved by the local administration, and binding him to submit to local regulations in the management of the affairs of the stream, and to keep his work in repair.

DETERMINING THE LEGAL HEIGHT OF DAMS.*

Dams for water-power purposes, and intended to hold the water at all times materially higher than the bed of the stream, are put in solidly from bank to bank, up to the least height at which it is necessary to hold the water for the purpose required, when the bank-lands above are sufficiently high, as they sometimes are, to be well above the flood plane as necessarily raised to higher levels by the effect of the dam.

But when these lands are not naturally high enough to admit of so high a flood plane, the top portion of the dam, for such height and length as may be necessary in each case, is made removable, automatically or otherwise, so as to admit free passage of floods through the weir thus opened, without their rising above a certain safe elevation in the reach above the work. These weirs can but seldom be dispensed with.

In the issuing of permits for the construction of water-power dams on non-navigable and non-floatable streams it was, until within the past fifteen or twenty years, the rule to restrict their heights so that the backset of waters would be confined to the limits of the lands

---

* See, particularly, De Passy, pp. 19, 23-25, 28, 51, and elsewhere; also, Dumont.

owned by the proprietors of the work, upon the theory that the bed of the stream was private property, and nothing could be done to affect it without liability for damage reclamation.

But the supreme court of France, and the council of state, have finally determined that "the fall of a stream of this character is not the property of the land proprietors, but that it enters into the class of things which, by the terms of article 714, Code Napoleon, do not belong to anybody, of which the use is common to all, and of which the enjoyment is regulated by the laws of police," and hence the administration sanctions works which cause water to be held back in the channels by properties above, so long as the *lands* and other works are not thereby injured.

In cases where it is necessary, in order to get head sufficient for the intended purpose, and at the same time guard against overflowings of land above, the administration is authorized to provide for the necessary levees on each side of the stream above the dam, to be built at the expense of the owner of the dam; all costs, charges, and damages being met by him.

The legal height having been determined for a dam, as a matter of record, and for reference at any time, a stone slab or shaft is firmly embedded at some convenient point, near at hand, where it can be conveniently got at, and so that its top surface is at the elevation of the dam's crest as authorized to be made.

Thus the officers of the administration, or any one else, may at any time test the fact as to whether or not the dam has been made higher than authorized.

This reference monument is an official record, and not to be displaced under pain of severe penalties, and the owner of the dam is responsible for its keeping.

CONSTRUCTION AND MAINTENANCE OF DAMS AND HEADWORKS.[*]

To provide for proper clearances of the beds of the stream above the dam, and to provide the means for permitting water in sufficient quantity to satisfy the rights which riparian proprietors below have under article 644 of the code, sluice-gates are put in all dams not built removable, at a point near or at the level of the natural stream bed.

Should it appear to a prefect in considering application for permission to put a structure for manufacturing purposes on a stream, that rights of riparian proprietors already utilized would be seriously injured by it, he has authority on this ground to refuse the permit—

---
[*] See, De Passy, pp. 24, 25, 51–54, and elsewhere.

the waters being already fully utilized and required for use under the code. The courts may order otherwise, however.

Dams established for irrigation, and made movable, cannot be used for power head purposes, and be kept closed all the time.

Dams for diversion of waters for irrigation must be removable down to the plane of the natural bed of the stream, for a length as great as the natural width of the stream between banks when cleaned out.

The movable portion must be composed of shutters which fall flat on to the bottom, of gates which may be raised above the flood plane, or of stakes ("needles") which can be taken out altogether.

The crest of the movable portion, like that of the fixed portion, must be adjusted to the plane of the legal height determined for the dam, and its sill must be established at the level of the bed of the stream when at its ordinary plane.

Scouring sluice-gates are not required in dams of this character, for the clearances above are effected by opening a portion of the dam itself down to the scouring plane.

Closable top weirs are also not required in dams of this kind, for a portion of the whole dam may be used for flood escapes.*

The dimensions and form of the head-gates of the canal, the elevation of their sills with respect to that of the top of the dam, the form and slope of the channels for a certain distance below, are regulated with the view of receiving the full flow of the water from the stream at low stage, when the division among claimants on the stream is made by giving each the full flow in turn at stated intervals, and, at the same time, to properly gauge a much smaller amount, when the division is made by apportioning the supply at once amongst a number, or all, according to their rights.

SECTION III.

REGULATION AND OPERATION—WORKS AND WATERS.

NECESSITY FOR REGULATIONS AND ADMINISTRATION.§

In the early years of the development of a new country, the necessity for guarding the common property of all the people is not felt. Each individual is intent on securing his own advantage, and all lose sight of those mutual interests which cannot be segregated and cut off in chunks as can lands and personal properties.

---

* Instructions, October 23, 1851.
§ See, Dumont, De Passy, Dalloz, De Buffon, and Malapert.

Water-courses and waters are, by nature, of the kind of property which no one can own, yet it has always been the idea in the early stages of the development of a people or a country, that each might use these common properties as he chose.

It was so in France. In the struggle for control of the navigable and raftable streams, which for centuries went on between the central government of the country and the nobles and the provincial governments, as I have already written, the small streams not raftable were left to the control and use of the riparian proprietors, the government maintaining a nominal and fitful supervision over them in the interest of public utility and the protection of navigation interests below.

Thus, customs became established which in course of time became crying abuses. So long as interests were few and water plenty in comparison to demand, and the stream banks were not much occupied, so long there was no pressing need of regulation other than that established by local custom and agreement.

But as time wore on, it was found that the courts were overwhelmed with water-right and other similar litigations. There was a perfect sea of trouble. The more decisions there were, the less were the people satisfied with the results.

It was found that water was used in the most extravagant and useless manner, and purposely or carelessly wasted by those who for long periods had enjoyed its control, while others equally well entitled to it originally, were forcibly deprived of a participation in its benefits.

The government was appealed to on all hands to make new laws, and indeed some legislation was brought about by this pressure and popular clamor.

But after awhile it was found that enunciation of principles, and formulation of general laws, and multiplying of rulings, without judicious and wise application of them according to local and ever varying circumstances, did not affect the desired ends.

RECOGNITION OF THE NECESSITY FOR ADMINISTRATION.

In the meanwhile it had become necessary for the government administration, in the interest of the public welfare, to interfere in these local quarrelings; and the salutary effect of these interferences became known and appreciated, seeing that regulation did away with litigation, and that the best was thus accomplished for all, with the advantages at command.

This led to the administration being called on in other cases, to establish regular rules and regulations on other streams; and so it has come about that on nearly all streams of any importance as sources of water supply for any purpose, or where their banks are built on, or where they run through municipalities, or are embanked to prevent floods, there are special regulations applicable to the cases which arise on them each.

It cannot be said that this system has been always acceptable to the people, or that it has not in places awakened violent opposition; for there has been opposition to administrative authority and control, and appeal taken to the courts. But the outcome is one of satisfaction with the principle on which rests the system, although, no doubt, the means of its application may not always be acceptable, and the results not always for the best.

Writers on these subjects of irrigation and drainage and the like, in France, with one accord unite in setting forth the necessity for a supervision of the affairs of water-courses.

Speaking of the diversion of water from, and construction of works in non-navigable streams, M.M. Dumont, being themselves advocates of the rights of riparian proprietors to control such streams, say:

"An unlimited freedom in this regard would be most dangerous. The privilege would be abused by some to the detriment of that of others, and of the public welfare. We must admit that if there were no regulations, every one could do as he chose, or use such quantities of water from the river as he willed, because of this privilege, and it would engender a veritable anarchy, and even lead to annihilation of law itself. There have been quarrels between irrigators and irrigators, and between these and factories, and these rival interests, not regulated, have been completely paralyzed, and all their advantages from a fair distribution of the water have been, in these cases, sacrificed.

"Therefore the exercise of the right of diversion from small running streams is and must be subordinated to certain conditions of general interest. In such matters the law cannot foresee all contingencies or regulate all cases, for what is good for one river is not good for another, and what is good for one season is not good for another.

"Hence all latitude and power is given to the administration in the exercise of its duty of improving and regulating the affairs of water-courses, to direct and manage them with the view to general utility, taking cognizance of the principles of irrigation."

"The courts themselves are required to conciliate the interests of agriculture with respect due to property, whenever litigation occurs between proprietors on these streams, to whom waters may be useful, and it has been expressly laid down for them that in every case they shall observe all particular and local regulations on the course and usage of water."

"The administrative regulating power, which is called upon to exercise so great an influence on the prosperity of agriculture, should rule over all water-courses, however small they are, even the waters of a brook fed by an intermittent spring."

De Buffon has written much in this same strain, and I have heretofore quoted from him, under a former heading, some strong sayings on the necessity for guards in carrying out regulations on the rivers. In another place, speaking of the bad condition into which channels have fallen for want of regulating their use, and the use of their banks, he says: "In the absence of rules of maintenance such a state of affairs is allowed to grow worse and worse during a number of years, and it will become intolerable, for a great extent of the riparian property will little by little lose its value, and other interests will be lost, because of conflicting and indeterminate claims."

"This is why a great number of localities are now suffering continually increasing injuries caused by the bad regime of these watercourses, and for that reason, in nearly every locality so affected, complaints are heard and demands made for the adopting of proper regulations and police measures to make an end of such a vexatious state of things. The superior administrative authority is continually solicited to favor the promotion of syndicates to act in concert with local administrations to insure the common good from the water-courses."

ADMINISTRATIVE AUTHORITY TO MAKE REGULATIONS.*

The authority of the central administration to make general and particular regulations governing the affairs of non-navigable streams is a power born of the natural necessity for regulation in the use of a property common to all the people, and of the recognized duty of government to foster the common interest, promote the general welfare, and protect the public rivers below, by establishing order in and imposing conditions on the diversion of waters from the tributaries above.

Hence, the origin of the authority of the central administration is not found in any laws or other enactments, but its duties are inferred from the laws and decrees relating to the subject and governing the action of the departmental administrative officers, and which have been already quoted.§

The duties with which we have most concern are those of "seeking and indicating the best way of utilizing the waters of all streams in irrigation," and others, which are of a police nature, in repression or prevention of individual license exercised to the detriment of common and public welfare.

The article 645 of the Code Napoleon modifies the power of the administration to interfere as between private rights to water on non-navigable streams, by relegating such questions to the courts. But these questions as to right being settled thus, or by long established usage, it remains for the administration to order matters from day to

---
* See, De Passy, Dumont, and Dalloz.
§ See, "Powers and Duties of Local Administrations."

day and year to year, in accordance with the basis thus established, and with the view of the public utility of the streams.

In cases where, under long established use, rights to definite quantities of water have become settled, the administration cannot do otherwise than recognize these rights, and establish regulations for the apportioning of the supply, in conformity with such claims.

Should all the rights be not already established by long use, the administration can only propose an apportionment, and, if this is not acceded to by the parties at interest, the case must be adjudicated before the proper courts, and then the administration establishes its regulations on the basis of the court's decree.

The administration has taken the authority to determine, however, the total volume of water which may be diverted for irrigation, as against the demands of navigation and manufacturing on the river below, and of deciding the dimensions of the headgates, etc., to take this water, and the periods of time during which it may be taken, and the court of appeals has sustained the acts of the administration in this respect, as being equitable and not in excess of authority.

When rights have been settled by long established usage, or by the courts, the prefects have the authority to establish regulations, in conformity with the schedule of rights thus fixed, defining the time, manner, etc., of use for each claimant, whether irrigator, manufacturer, or commune, and according to existing circumstances.*

But, if no settled rights exist, regulations always emanate from the council of state in general assembly, for to the sovereign authority belongs the right to settle matters so nearly affecting the general interest.

"From these principles as to authority, it follows that in the absence of long established and recognized custom and local usage, and in cases where it becomes necessary in the general interest to modify such practice, there is no other provision for a division of water in this class of water-courses among the several users, but a decree emanating from the council of state in general assembly."—[De Passy.

PRINCIPLES ADHERED TO IN MAKING REGULATIONS.§

In cases where a division of water is to be made between agriculture and industrial pursuits, the points to be fixed are of two kinds—those special to each particular case and those common to the whole set of cases. The special points are the following:

First—During what periods is it necessary to have water for irrigation. First, for the spring waterings, and, second, for the watering

---
* Decree of April 13, 1861.
§ See, De Passy, Dalloz, and Dumont.

of summer crops; and on what days, and at what hours during each of these periods, will it be necessary to have the water.

Second—In what divisions of the stream do groups of distinct and separate interests lie; what is the extent of interest in each division; what proportion of the whole available water supply reckoned in days and hours will be required in each division; at what times will each division demand its proportion; and what is the constant demand in each division for water for domestic purposes.

The general points are as follows: The waters set aside for manufacturing power purposes, are after use or when not used, accorded to irrigation without regulation, unless the considerable number of interests on the stream below makes a schedule necessary to preserve order in division.

The gauging, rating, guarding, and operation of the headgates of canals and sluices, and of the weirs and open ways of dams, is the subject of a general regulation.

The making of a general schedule for division of waters, and of a special card therefrom for guidance in the use of waters at each manufactory and by each irrigation canal, is the subject of a general regulation for the stream.

The reservation of waters for purposes other than those specified in the schedule, in the interest of the public generally and parties using water from the stream for other purposes than irrigation and power, is the subject of a general regulation for the stream.

The distribution in irrigation by the irrigators themselves, of the waters allotted to them in each case, and provision for citing them before the courts to have their matters of dispute settled, under article 645 of the civil code, so that water be not wasted while they are quarreling, is the subject of a general regulation for the stream.

A type of public administrative rulings for a division of water between agriculture and industrial pursuits, is the decree dated July 2, 1872, relative to the river Furè in the Department of Isère, hereinafter given, under the head of "Regulations of Irrigation."

GENERAL RULES AS TO DIVISION OF WATERS.[*]

In the issue of permits to construct dams for irrigation in watercourses of this class, a special obligation is imposed on the owner of the work, that the water passage shall always remain open, and thus a free flow of the stream on its natural bed be assured, except when the water is being diverted into the canal as provided for in the schedule of division.

---

[*] See, De Passy; also, Dumont, and Dalloz.

This provision is necessary to guard against floodings above the dams, and to insure a fair distribution of the waters according to the schedules, and to allow the stream to keep itself clear from deposits caused by the dams when closed; and the necessity for it has been made glaringly apparent by a long and disastrous experience with dams not provided with open ways.

In cases where the water volume in the stream to be divided is sufficiently large to admit of all claimants receiving sufficiently large irrigating or power heads at once, the schedule is made on this basis; but if the supply is not sufficient for this purpose, the system of "turns" by the day, week, or hour is adopted, and the schedule so arranged as to accommodate as many as possible with the supply under this arrangement.

The system of turns is preferred by the administration as well as the irrigators on one account, and that is, because the supervision has then only to be directed to fixing the time for opening and closing the headgates and dams and not also to the regulation of the amount they shall be opened.

But this system has the disadvantage often of not allowing the waterings to be made when the crops most need it.

The administration, in making schedules for divisions of water, is governed by ancient local custom, probable water supply, and as far as possible by the necessities of each individual water-right holder; so that in reality it only acts as a disinterested third party apportioning a common benefit, as far as possible to suit desires of the parties most at interest, and reserving and caring for the rights of other parties at interest, much scattered and not otherwise represented.

In authorizing the construction of a new work by a party having a riparian right to water, the prefect, representing the administration, if there are well established general rules or customs governing water division on the stream, inserts a clause to the effect that the new work is to be used in conformity to such rules as carried out by the administration or the consumers amicably amongst themselves.

In the absence of ancient rules or customs the prefectorial order is limited to authorizing the construction of the work, leaving for the future the determination in the general interest, of conditions under which the new work is to be used, if it should become necessary so to do, or, if this becomes necessary also, awaiting the action of the courts in determining the relative rights of the parties at interest.

Thus, questions relating to the actual right to water, the relative extent of each claim to water, the right to partly or wholly support a dam on another's land, the right of way to conduct water over another's land, the point at which drainage waters shall be returned to the streams whence the head is derived, and, in a word, all questions relating to each individual claim are, if necessary, first to be adjudicated by the courts, and the administration bases its regulations on these decrees.

REGULATIONS OF IRRIGATION—DIVISION OF WATERS BETWEEN CLAIMANTS.[*]

As a practical example of an administrative measure regulating the division of waters between agriculture and manufacturing and other industries, the following decree of the president of the republic, dated July 2, 1872, is given in full.

It will be understood, of course, that the waters, except when being used, as specified, in irrigation, are to remain in the channel for power generation at the dams devoted to other purposes than irrigation.

"The president of the French republic, in view of the decree of the 5th May, 1865, declaring to be of public utility the works for the management of the lake of Paladru, intended to supply, for all time, to the river of the Furè, the volume of water sufficient for the necessities of irrigation of the river meadows, and the working of numerous manufactories which exist on this river.

In view of the reports of the engineers of the department of the Isère, relative to the measures to be taken to do away with the abuses proceeding from the absence of schedules regulating the use of water.

In view of the documents of the two inquiries opened by prefectorial judgments of 4th November, 1867, and 18th May, 1871.

In view of the opinion of the commission of the syndicate of the Fure, in date of 10th October, 1870.

In view of the uniformity of plan of the valley of the Fure, and the proposition of the proprietors of the irrigated meadows.

In view of the reports of the engineers in date of 16th February and 31st May, 1870, 19th November, 1871, and 29th February, 1872.

In view of the opinion of the prefect in date of 13th March, 1872.

In view of the opinion of the general council of bridges and roads in date of 27th March, 1872.

In view of the laws of 12–20 August, 1790, 6th October, 1791, and the judgment of the government of 19th Ventose, year 6, the decree of decentralization of 13th April, 1861.

And the temporary commission, charged with replacing the council of state, being heard, renders judgment as follows:

Article 1. From 1st March to the 1st September, each year, the meadows which have the right to the waters of the Furè, on the territory of the seven communities of Charavines, Apprieu, Saint Blaise de Buis, Beaumont, Rives, Renage, and Tullins, will be irrigated once a week.

First—The meadows included between the source of the river and the dam of headworks of the furnaces of Riviere, a point situated at 2028.50 metres down stream from the bridge of the departmental road No. 7, from Sunday at one o'clock in the morning till Sunday at half-past seven in the evening.

---

[*] See, De Passy, appendix No. 1.

Second—The meadows included between the dam or headworks of the furnace of Riviere and the mouth of the stream of Réaumont, in the Fure, from Saturday at nine o'clock in the evening till Sunday at half-past seven in the evening, to wit: from Saturday at nine o'clock in the evening till Sunday at one o'clock in the morning, with the total discharge of the stream, and during the remainder of the time, with the product of the waters of filtration, proceeding from irrigations up stream, and that of the tributaries which fall in this part of the bed of the Fure.

Third—The meadows included from the mouth of the stream of Réaumont, and the end of the course of the Fure, from Saturday at six in the evening, till Sunday at half-past seven in the evening, to wit: from Saturday at six in the evening, till Saturday at nine in the evening, with the total discharge of the water-course, and during the remainder of the time, with the product of the waters of filtration, proceeding from the irrigations up stream, and that of the stream of Réaumont, as well as the tributaries which fall in this part of the bed of the Fure.

Article 2. The proprietors of the meadows will have, nevertheless, the power of practicing supplementary irrigations, when there are superfluous waters, that is to say, when the manufactories are working regularly, and the river affords an excess of discharge, it may be passing across the sluices of discharge, raised for this purpose by the manufacturers, or it may be by accidental overflowing above the weir.

The irrigators can open their headgates, but on condition of closing them, as soon as the water of the river will have descended to the legal level of the dams, the sluices of discharge being closed.

Article 3. Outside of the fixed hours for irrigation, by article 1, and except the case of use of superfluous waters, under the conditions provided by article 2, the sluices of the irrigation dams existing, it may be on the Fure, it may be on the millponds taken from this river, will have to be completely raised above the level of flood waters, and the sluices of the headworks will remain tightly closed.

The proprietors, having, in virtue of titles legally recognized, a right to a continuous small stream of water, it may be for their domestic uses, it may be for feeding their retting pits, will be able at all times to preserve in their respective headgates the openings necessary to receive the continuous volume of which they have the right of enjoyment.

Article 4. In the regulating schedules for the works intended to assure the irrigation of the meadows, and the régime of the manufactories, the prefect will fix the conditions, which he will judge necessary with the purpose of maintaining the division of the waters made by the present decree.

Article 5. The rights of outside parties are and continue expressly reserved.

Article 6. The irrigators will arrange between themselves for dividing the waters placed at their disposition, and will carry all disputes which may arise from said division of waters, before the competent authority.

Article 7. The minister of public works is charged with the execution of the present decree.

As a practical example of the regulations of police of non-navigable water-courses, the following formula promulgated in 1878, as a circular, to the local administrative officers, by the minister of public works, is presented.

It is explained that this is intended as an outline to be followed by the prefects in getting up general regulations for the streams in their departments.

Obligations of the Riparian Owners.—Riparian owners are to lop off and remove all trees, bushes, and stumps which might form an obstruction on the banks of the water-course, and all the branches, which, touching the water, might impede the flow.

Silt Accumulations.—Riparian owners are obliged to receive on their lands the materials coming from the cleansings of the channel, and to remove the deposits which would injure the free flow of the waters.

Passage of Riparian Properties.—The riparian owners are obliged to give free passage over their lands, from the rising to the setting of the sun, to the officers and their agents in the discharge of their duties, as well as to the foremen and workmen charged with cleansings of the streams.

These persons cannot, however, use the right of passage over closed lands, except after having previously notified the owners.

In case of refusal they will require the assistance of the mayor of the community. They will be responsible, besides, for all damage or injury committed by them or their workmen.

Construction.—Every proprietor who wishes to make a structure, or a change in any structure, upon the water-course, or adjoining it, must submit to the prefect the plan of the work he proposes to adopt.

In the two months which follow the deposit of this communication, the prefect, after having taken the advice of the engineers, will make known to the petitioner if the projected works would appear to injure the free passage of the waters, and if, in consequence, the administration is opposed to their execution.

After this delay, if he has not received any response, the petitioner can go ahead, without, however, prejudicing the rights of third parties, and those of the administration.

No dam, plantation, permanent or temporary work, of a nature to modify the régime of the waters, may be established or repaired on a water-course without the authorization of the prefect.

It is forbidden to make ditches in the banks, or practice any other means of derivation, without having first obtained the permission of the prefect.

Obligations of Manufacturers and Users of Dams.—The weirs and sluices of discharge will always be maintained open, and it is expressly forbidden to place anything on them for the purpose of raising them.

In default of an official ruling which fixes the legal height of the dam, the waters are not to pass over the upper part of the weir, or from the sluice of discharge with a head of pressure if there is no weir.

Manufacturers and users of the dams will be responsible for the super-elevation of the waters, as well as when the discharge sluices are not raised to their full height.

---

* See, *Les Annales des Ponts et Chaussées*, Vol. CXXXIX, p. 1112.

The manufacturers and users of the dams will be obliged to open their sluices for the execution of the works of cleansing, during the hours and days which will be fixed by the prefectorial decrees made upon the advice of the engineers.

Deposits and Injurious Waters.—It is forbidden to make any deposits in the bed of a stream or to allow infectious or injurious waters to drain into it.

The interdiction made by article 17 of the decree above vised, 10th August, 1875, of fishing in the parts of streams of which the level would have been temporarily lowered, it may be by conducting the cleansings or any kind of works, it may be on account of the stoppage of the manufactories, is reaffirmed.

River Guards.—There will be river guards organized and specially charged with putting in operation the present rules, provided that all the interested parties or any certain number of them, have made an engagement among themselves to assure the payment of these agents, under the subventions which would be furnished by the state, the department, or the communities.

These agents will be commissioned by the sub-prefect, and will be sworn before the tribunal of the district.

Infringements of the rulings of the present law will be proven by means of statements drawn up by a river-guard, or by any other agent of authority who has qualified for this purpose.

These statements will be affirmed within three days of their date, before the mayor or justice of the peace, either at the residence of the agent, or in the place of the offense.

They will be vised for stamps and registered fee, in the space of four days after the affirmation, and referred to the competent jurisdiction.

A copy of each statement will be remitted by the agent who will have drawn it up, to the mayor of the commune, who will certify to it and send it to the infringer, with the summons, if necessary, to cease immediately from damage.

The present regulation will be published and posted throughout the extent of the department.

Copies of it will be addressed to the engineer-in-chief, to the sub-prefects and the mayors charged, each one in that which concerns his business of overseeing and assuring the execution of the prescribed rulings.

AUTHORITIES FOR CHAPTER IV.

In the preparation of this chapter, I have consulted and compared the following-named authorities:

Dumont.—[Work cited as an authority for Chapter II (French)]. See Book II, Chapters II, III, and IV.

De Passy.—[Work cited as an authority for Chapter II (French)]. See Chapter I, pp. 14–130, and supplement, pp. 297–334.

Malapert.—[Work cited as an authority for Chapter II (French)]. See headings, "Actual Republic," "Engineers," "Water-Courses."

De Buffon.—[Work cited as an authority for Chapter II (French)]. See Vol. 2, Part II, Sec. I, pp. 1–106.

Dalloz.—[Work cited as an authority for Chapter II (French)]. See Vol. XIX, "Waters," Chapters IV, IX, and X; also, Vol. XL, title "Servitudes."

Les Annales des Ponts et Chaussées.—[Work cited as an authority for Chapter II (French)]. See, particularly, Vol. CXXXIX, p. 1112 et seq., and also the late volumes of "Laws and Decrees."

Civil Code.—[Works cited as authority for Chapter II.]

CHAPTER V.—FRANCE[4];

RIGHT OF PROPERTY IN SPRINGS, AND RIGHTS TO THE USE OF SPRING WATERS.

SECTION I.—*Ownership and Control of Springs.*
    Absolute Ownership.
    The Opposing Doctrine.
    The Settled Principle.

SECTION II.—*Acquired Rights to Spring Waters.*
    Public Use of Springs; Populations.
    Private Use—By Title; Prescription.
    Servitude Resulting from Dividing Estates.

SECTION III.—*Rights of Drainage and other Rights.*
    Natural Right of Drainage—Civil Code.
    The Right to Dig or Bore for Water.

SECTION I.

OWNERSHIP AND CONTROL OF SPRINGS.

ABSOLUTE OWNERSHIP.[*]

The matter of the ownership and control of springs has been one full of contention in France. But it is now well settled by the provisions of the code, and the decisions under it. Article 641 of the civil code says: "He who possesses a spring within his field may make use of it at his pleasure."

It follows from this that, "a spring is the exclusive property of him on whose land it rises, and is used in an absolute manner like the land itself. The owner may lead its waters over his land, change their course, collect them in ponds and reservoirs, cause them to be absorbed by the ground, or even suppress the spring itself, and his neighbors will protest in vain against being deprived of them."—[Dumont, § 127.

The code, however, defines certain circumstances under which this control of springs is limited and qualified; the causes being— the necessities of communities for water for domestic purposes, the necessities of the State for water for purposes of navigation, the

---

[*] See, Civil Code, Articles 641, 642, 643; Dumont, §§ 127-129; De Passy, p. 21, and elsewhere; Dalloz, Vol. XXXVIII, p. 217, and Vol. XIX, p. 398; also, Proudhon.

rights which persons other than the owners of springs may have acquired to the use of their waters by purchase or by prescription.

The injunction laid upon the courts by article 645 of the civil code, which commands that "if a dispute arise between the proprietors to whom such waters may be useful," they, the courts, "in pronouncing judgment, must reconcile the interests of agriculture with the respect due to property," applies only to waters mentioned in article 644, namely, those of non-navigable and non-raftable streams, on the use of whose waters, in favor of riparian lands, a servitude is laid, and does not apply to the waters of springs.

Hence, the courts have not the power to partition the waters of springs between the proprietors to whom they may be useful, as in the case of waters of small streams, and the administrative department has never attempted it as a regulation.

THE OPPOSING DOCTRINE.*

This doctrine has been strongly opposed in France, however, and there are writers, and some decisions, which hold that the principle of compromise and judicial control, embodied in article 641, was meant for application in the case of springs, as well as in the case of small water-courses, and that hence the courts can, in the interests of agriculture in general, and for the benefit of local agriculturists in particular, prevent the unnecessary wasteful or selfish use of spring waters, as well as those of a stream by an owner on its banks, and compel a division of the water with owners of adjacent lands, if there is really more water than is necessary for the lands containing the source, and for the legitimate necessities of the proprietor.

THE SETTLED PRINCIPLE. ₰

The ownership and control of springs is so complete and absolute that, so long as the waters remain within the property where they rise, even though used for manufacturing, power purposes, or otherwise, the administration, which has such extended authority in the regulation of the use of waters under other circumstances, can do nothing to interfere with the proprietor's use of the spring waters, "even though they be in sufficient volume to form a veritable water-course." [De Passy, p. 21.

"With regard to springs which rise on the lands of an estate * * * they belong to the proprietor of the lands themselves. * * * The proprietor, then, disposes entirely of the spring, saving the rights which may have been acquired against him, and saving the sacrifices

---

* See, Dumont, ₰ 128.
₰ See, De Passy, p. 21, and elsewhere.

which the public interest may exact to the detriment of his right." [Dalloz, Vol. 38, p. 217.

But if spring waters be led across or into property other than that containing the source, no matter though the using be for the benefit of the owner of the source, or for whatever purpose, such stream is subject to regulation, as in the case of others.

SECTION II.

ACQUIRED RIGHTS TO SPRING WATERS.

PUBLIC USE OF SPRINGS—POPULATIONS.*

Private interests must always be subordinate to public interests, however, and on this account the owner of a spring cannot change the course of its waters when they furnish the necessary supply to the inhabitants of a commune, village, or hamlet. "The legislature has always held in view the personal necessities of people rather than the requirements of agriculture, as necessary to the moral well-being of the nation."—[Dumont, § 130.

This servitude is sometimes burdensome upon the proprietor of an estate who may desire to divert the waters of his spring to some purpose useful to himself, and, hence, he has the right to claim payment from the community, unless the inhabitants have, by use for a due length of time, a prescriptive right to the water. "The amount of the indemnity is determined by the courts, who take into consideration the degree of injury proved by the proprietor, rather than the advantages reaped by the commune, village, or hamlet."—[Dumont, § 130.

Government can also take possession of springs to feed canals for navigation, but on condition that it pay a just indemnity, as adjudged by the courts, and in conformity to the law for the condemnation of private property to public use. §

"It has been decided that a spring existing in the land of an individual is presumed to be the property of a community of people when this community has had the continual use of it from time immemorial, for domestic and community purposes."—[Dalloz, Vol. 19, p. 217.

PRIVATE USE—BY TITLE: PRESCRIPTION.†

The absolute right of ownership in a spring is also modified by

---
* See, Dumont, §§ 130, 131: Dalloz, Vol. XXXVIII, p. 217; Proudhon, p. 4.
§ Law of May 3, 1841.
† Dumont, §§ 132, 133, 134, 139½; Dalloz, Vol. XL, title "Servitude"; Civil Code, arts. 688, 689, 690, 691.

purchased titles, by prescription, and by the servitude set up by the division of an estate containing a spring.

A purchased right to the use of the waters of a spring is evidenced by a deed or record from the owner or former owner of the spring. In cases of uncertain meaning to such documents, the Courts adhere to the presumption that the owner of the spring did not mean to restrict his own use of the waters in the fullest extent necessary for his purposes, but only to give the grantee the right to control the waters at any time found running in the channel below.

"The right most commonly ceded to a third party, upon a spring, is that of drawing water, or that of leading water away from it. The servitude thus accorded is regulated by the principles of conventional servitudes.

"The concession of a right of leading out water does not prevent the proprietor from himself using the water of the spring for the wants of his property, but he cannot change the cultivation of his property in such a way as to absorb a greater quantity of water than he was using at the moment of the concession.

"He who has ceded upon his spring a right of leading out water, can cede another to another person, without the consent of the first cessionary, provided always that the waters thus divided amongst several cessionaries can still suffice for the wants of each; otherwise, the consent of the first cessionary will be needed.

"The owner of a property to which the servitude of leading out water is due, cannot, without the consent of him who owns the property which owes it, concede it to a third party, nor even use the water for another property, or for another part of the property.

"One can acquire a servitude of leading out water on a higher property, from which it is separated by an immediate property or by a public road. In the latter case an authorization is necessary. There is a servitude of aqueduct on the intermediate property, and a servitude of leading out water on the higher property. The proprietor of the intermediate property cannot serve himself with the water which passes through his land, without the consent of his two neighbors who have treated for the servitude of the water-right."—[Dalloz, Vol. 40, word "Servitude."

A prescriptive right to the use of the waters of a spring is "acquired by an uninterrupted enjoyment of them during the space of thirty years; to be computed from the moment at which the proprietor of the lower field has made and completed the works apparently designed to facilitate the fall and course of the water within his property."*

The courts hold that the essential points to be established in proving this servitude are:

(1) That the works have been established in a permanent manner, (2) and maintained for thirty years, (3) in a manner to constitute an adverse possession of the water to that of the owner of the spring, and, hence, in consequence of the last condition, that these works be

---

* Civil Code, Art. 642; see, also, Articles 688, 689, 690, 691.

attached to the tract wherein the water rises. "This last condition is not written in the law, but it is the meaning of it, and this point, which has been the subject of lively debate, is at present sanctioned by jurisprudence."—[Dumont, § 134.

"The second exception to absolute ownership in a spring, on the part of him who has it on his property, consists in the prescription which can be acquired of the right to use the water of this spring.

"Prescription in this case can only be acquired by uninterrupted enjoyment, during thirty years, counting from the moment in which the proprietor of the lower land has made and terminated visible works destined to facilitate the fall and flow of the water on his property.

"We will remark at first that the prescription does not apply to a simple right of drawing water; for that is a discontinuous servitude, and servitudes of that description are not acquired by prescription. It would be different with a servitude of this class which would have been acquired by possession before the publication of the civil code.

"The general principles of prescription receive here their application.

"Moreover article 642 establishes special rules of which the accomplishment is necessary in order that the servitude may be acquired by prescription.

"It is necessary in the first place that there may be works. In vain the higher proprietor would have allowed the lower property to enjoy peaceably and publicly the use of the waters; this would only be a simple tolerance which could not constitute a right."—[Dalloz, Vol. 40, word "Servitude."

THE SERVITUDE RESULTING FROM DIVIDING ESTATES.[*]

There are cases wherein lower and other proprietors hold the right to use the waters of a spring otherwise than by purchase or prescriptive use for thirty years.

Thus when an estate containing a spring has been subdivided amongst heirs, after having been held by one proprietor, and the waters used to the benefit of the lower lands, so as to result in a servitude, by the owner of all, the owners in common and co-heritors of the upper and lower part of the estate share the use of the waters after the division of the lands.

This servitude results from article 692, civil code, which is as follows: "The declaration of the father of a family is equivalent to a deed as regards continual and apparent servitudes."

The rights of ownership and use of a spring may be restricted, but not annihilated by the servitude above named, and it rests with the courts to conciliate the several interests in such cases.

The rights above described, acquired by prescription and the "servi-

---

[*] See, Dumont, §§ 134, 136, 137; Civil Code, Arts. 688, 689, 692.

tude of the father of a family," do not constitute property rights, either in the spring or its waters, but simple rights to the use of some portion of the water, according to the facts in each case.

Thus, the possessor of the lands in favor of which such rights have accrued, can not take water at such times, and in such manner, and in such quantity as seems best to him. "Conciliating the right to use with the rights of the owner of the spring, the courts can decide that in the future he does not use the water, but according to a measure which, in default of an amicable agreement, will be regulated by the courts, by experts." This duty of experting usually falls to the engineers of the administration in charge of streams.

SECTION III.

DRAINAGE AND OTHER RIGHTS.

NATURAL RIGHT OF DRAINAGE.*

Article 640 of the civil code reads as follows: "Inferior lands are subjected as regards those which lie higher, to receive the waters which flow naturally therefrom, to which the hand of man has not contributed.

"The proprietor of the lower ground cannot raise a bank which will prevent such flowing.

"The superior proprietor of the higher lands cannot do anything to increase the servitude of the lower."

Under this article, drainage waters from springs must be permitted to flow as they would naturally flow on to lower lands.

If the ordinary clearing or cultivation of a field, or excavation for ordinary purposes other than those of developing a flow of ground water, causes an increase in the flow of a spring, or the breaking out of a new one, these waters must be allowed to drain away as though naturally started.

The owner of a lower estate cannot, however, without due indemnity, be made to suffer the passage over his lands of waters caused to flow by excavations made for the purpose of getting a flow of water, or where it is well known a harmful flow will result, or by artesian borings.

THE RIGHT TO DIG OR BORE FOR WATER.§

Article 552, civil code, reads as follows:

"Property in the soil imports property above and beneath.

---

* See, Dumont, § 129; also, Dalloz, title "Servitude."
§ See, Dumont, §§ 138, 139; and Dalloz, Vols. XIX and XL, words cited.

"The proprietor may make above," etc. * * * *

"He may make beneath, all structures and excavations which he shall judge convenient, and draw from such excavations all the products which they are capable of furnishing, saving the restrictions resulting from the laws and statutes relating to mines, and from the laws and regulations of police."

In consequence of this article, ownership of land carries with it all above and under the soil.

The application of this principle authorizes the land owner to make on his land any works or excavations he deems expedient for his purposes, even though they result in the cutting of subterranean veins of water that feed a spring rising upon the lands of a lower proprietor.

"The court of appeals has even extended this privilege to cases where such excavations would damage mineral water establishments belonging to the State, and it refused the administrative authority of the mayor of Vichy the power to render decrees to forbid such excavations." * * *

"The council of state has also sanctioned the same principle in a similar case."—[Dumont, §138.

This natural privilege may be forfeited by agreement amongst proprietors, so that one estate be bound not to excavate to the detriment of waters or springs naturally rising on another.

AUTHORITIES FOR CHAPTER V.

In the preparation of this chapter I have consulted and compared the following named authorities:

Dumont.—[Work cited as an authority for Chapter II (French).] See, Book II, Chap. IV, pp. 209–225.

De Passy.—[Work cited as an authority for Chapter II (French).] See, pp. 21, 22, and elsewhere.

Dalloz.—[Works cited as authority for Chapter II (French).] See, Vol. XIX, title " Waters," p. 276, and elsewhere, and Vol. XXXVIII, title " Property," p. 217, and elsewhere, and Vol. XL, title " Servitudes."

Proudhon.—[Work cited as an authority for Chapter II (French).] See, p. 4, and elsewhere.

Civil Code.—[Works cited as authority for Chapter II.] See, particularly, Arts. 552, 640, 641, 642, 643, 688, 689, 690, 691, 692.

CHAPTER VI.—FRANCE[5];

THE RIGHT OF WAY TO CONDUCT WATER AND THE RIGHT TO ABUT A DAM.

SECTION I.—*Rights for Works of Public Importance.*
Condemnation for Works of Public Utility.
Way for Main and Secondary Works.
The Laws of 1836 and 1841.

SECTION II.—*Rights for Private Water-Ways.*
Servitude of Right of Way; Law of 1845.
Servitude of Right for a Dam; Law of 1847.
Application of these Laws.

SECTION I.

RIGHTS FOR WORKS OF PUBLIC IMPORTANCE.

CONDEMNATION FOR WORKS OF PUBLIC UTILITY.*

The right to land, or to occupy land upon which to locate a canal or other water conduit, with its accessory works and structures, is, according to circumstances, obtained in France either by acquiring title to the strip of land itself, or as a servitude or right of occupation and use for the specified purpose.

In acquiring *title* to lands for the location of works, the mode of amicable private purchase is always open, and is the only means of attaining this desired end until the project shall have been declared and recognized by law or decree as being of public utility or importance, when the properties may be condemned as for public use.

This process of condemnation is carried on under laws of 1836, regarding local roads, and of 1841, regarding expropriation for causes of public utility.

Expropriation, or condemnation of private properties for works of public utility, is accomplished through the action of the courts, which, however, can only order the condemnation after the declaration of public utility has been made, for each case, (1) in the special law or ordinance which authorizes the execution of the works for which the

---

*See, particularly, *Les Annales des Ponts et Chaussées*, Vol. XX, pp. 203-217, and Vol. XII, p. 328 *et seq.;* also, Dumont.

expropriation is required, (2) in the decree of the prefect which designates the localities of the tracts on which the works are to be placed (when this designation is not contained in the law or ordinance), and, (3) in the final decree in which the prefect designates the particular pieces of property, according to ownership, metes and bounds, which it is necessary to condemn; and such condemnation can only be made after due hearing of interested parties, and in conformity to process of law.

Great public works, such as national roads, railroads, basins and docks, canals, and the canalization of rivers, whether enterprises of the state, of departments, communities, or of particular companies, whether toll is to be charged in any way or not, or whether a subsidy of treasure is to be granted or not, or whether any part of the public domain is to be used or not, can only be executed by virtue of a special law, which can be passed only after an administrative inquiry has demonstrated the feasibility and desirability of the work, and a report has recommended it.

A central administrative ordinance is sufficient to authorize the execution of departmental routes, that of canals and branch railroads less than 20,000 metres in length, and of bridges and other works of less importance; but such ordinance must also be preceded by due inquiry, examination, and report on the project, in conformity with regulations formulated by the central administration.

With respect to the administrative and legal forms to be followed in the condemnation of properties for works declared to be of public utility, this law goes into minute details at great length, expressly defining and prescribing each step to be taken, under the following general headings: Administrative measures of inquiry preceding condemnation; effect of condemnation on mortgages and other similar rights; the rule of indemnification; the payment of indemnities; contracts of sale; and others not at all necessary to enumerate.

WAY FOR MAIN AND SECONDARY WORKS.

From the first part of this long law, it appears that wherever it is proposed to condemn property for purposes of public works, such as for right of way for a canal, there must first be a report from the government engineers defining or recommending the proposed route, and showing the lands, etc., proposed to be taken in each commune or community. This plan is posted at the local mayoralty house, and advertised for inspection of all concerned. Thereafter, an inquiry is held by a commission to hear all objections, criticisms, or suggestions of change. On the result of the report of this commission, with the evidence annexed, the prefect designates the route to be taken and defines the properties it will be necessary to take for the work. Should it be necessary from the report of the commission to modify the plans proposed for the works, the subject must be referred to the central administration, and the prefect awaits its decision. The properties being thus defined, the question becomes one for the courts, according to the provisions of the law which follow under the headings already given.

In accordance with this law, whenever a canal enterprise of importance is to be authorized, so that the projectors may have the right of condemning private property for right of way or other necessary purposes of the work, there is a special law passed which declares the proposed work to be one of public utility, and entitled to the benefits of the provisions of the laws providing for the condemnation of private properties for public use.

This method of acquiring right of way for great works of public importance is of ancient origin in French legislation, for although the special laws cited are of comparatively recent date, they are founded on and are elaborations of others preceding them.

These provisions, however, applied only to rights of way for main works—those which could be recognized as being of public utility; and until 1845 there was no method, except by amicable private purchase, to acquire rights of way for the minor distributing ditches of great canal systems, nor was there any possibility of a private individual or of any organization acquiring a right to conduct water over lands against the will of the owner of the lands, until the work had been officially examined and declared to be of public utility as above explained.

SECTION II.

RIGHTS FOR PRIVATE WATER-WAYS.

SERVITUDE OF RIGHT OF WAY.*

The passage of the law of 1841, on the condemnation of private properties for purposes of public utility, which was really in this respect a re-enunciation of laws already existing, brought the right of way question to a head, so that in 1843 a proposition was introduced in the chamber of deputies, for a law declaring that *all* irrigation works constructed by companies or *individuals* should be declared to

---

* See, particularly, Dumont, Book II, Chap. V, and *Les Annales des Ponts et Chaussées*, Laws and Decrees, 1845 and 1847; also, De Passy.

be of public use according to the forms of the law of 1841. And this, in turn, caused the introduction of another proposition for a law of dispossession for right of way in favor of *all proprietors*, whether owners of bank lands or not, who wanted to use water for the irrigation of their estates.

It was pointed out at the time that one of these propositions was opposed to the principle of the fundamental law of the country—that private property could only be condemned for public and not for private use; and that the other proposition was opposed to the well established exclusive right of riparian owners to waters of non-navigable and non-raftable streams. The whole question of a draft of a law as a substitute for these was then referred to a commission, and this commission reported, and the chambers, after a long consideration, passed the law, which here follows:

Law upon the Right of Way for a Canal—Passed twenty-ninth of April, 1845.

Article 1. Every proprietor who may wish to be served for the irrigation of his property with the natural or artificial * waters of which he has the right to dispose, can obtain the passage for these waters over intermediate lands by previously paying a just indemnity.

There are excepted from this servitude houses, pleasure grounds, gardens, parks, and inclosures belonging to dwellings.

Article 2. The proprietors of lower lands will have to receive the waters which percolate from lands thus irrigated; being indemnified, however, if damaged.

Houses, pleasure grounds, gardens, parks, and inclosures belonging to dwellings will be equally excepted from this servitude.

Article 3. The same right of passage over intervening lands will have to be accorded to the proprietor of a property submerged in whole or in part, for the purposes of drainage.

Article 4. The questions to which the establishment of this service will give rise, the fixing of alignment of the water conduit, of its dimensions, and of its form, and the indemnities due—it may be to the proprietor of the land traversed, it may be to that of the property which will receive the drainage waters—will have to be taken before the courts, which in pronouncing on them will have to conciliate the interest of the enterprise with the respect due to property.

It will be tried before the tribunal in a summary manner, and if a question for experting, it will only be necessary to name one single expert.

Article 5. There will be nothing detracted by the present provisions from the laws which regulate the police of waters.

The consideration of this law on its passage gave rise to long and stormy debates in the chambers of the legislature, in which it was attacked on about the same grounds as those previously referred to

---
* "Artificial" waters: those drawn from deep wells or otherwise brought to the surface of the ground artificially.

the commission. A synopsis, with extracts from the speeches at length of these debaters, is given by M.M. Dumont, and as the result of their consideration of the subject the following conclusions are drawn:

First—That the law had for its sole object the establishment of a legal servitude to be laid on property in obtaining a right of way to conduct across it such waters as one has the right to dispose of.

Second—That it leaves intact all the points of the laws and decisions preceding it and relative to the ownership and police of waters.

These conclusions have since been repeatedly verified by decisions of the courts of highest resort.

The nature of this servitude and the spirit in which it was advocated may be well understood from the following: In the course of the debate the judge advocate said, "the judicial power can according to the case, grant or refuse the servitude, as it is or is not justified by real irrigation interest;" and commenting on this and other paragraphs M.M. Dumont say:

"It is without doubt that the courts are not obliged to grant the servitude of passage every time it is demanded; on the contrary the law imposes on them the duty to estimate the degree of usefulness it has, to balance this usefulness with the injury that the digging of the canal might cause to properties, to examine if the water proposed to be diverted has not already an equally beneficial application, and, finally, to consider all the circumstances of the case."

"The servitude is created for the benefit of lands *for irrigation*, and not for conducting water for ornamental or any other purpose, and the courts will refuse to allow its application for any other purpose than those of the irrigator."

SERVITUDE OF RIGHT TO ABUT A DAM.[*]

The passage of this right of way law went far to clear away the difficulties attending the establishment of private irrigation works by riparian proprietors on the non-navigable streams, and those who had obtained water concessions on public streams from the administration, and those who owned the water of springs. But a great difficulty yet remaining was that of acquiring the right to construct a dam against the bank of another riparian proprietor. One might own one bank of a stream yet could not build a dam in it to divert water on to his own land, should the owner of the opposite bank object to the end of the dam being rested against his land. Or one might have right of way to conduct water, but not right to put a dam in a stream to divert it, because the bank owners objected, and this, too, when the administration may have approved the project.

---
[*] See, De Passy, and Dumont.

This condition of affairs led to great conflicts, and these resulted in the passage of the following law:

Law Upon the Right to Abut a Dam—July 11, 1847.*

Article 1. Every proprietor who will wish to be served for the irrigation of his property with the natural or artificial waters of which he has the right to dispose, will be able to obtain the privilege of supporting upon the property of the opposite bank-owner the works necessary for its taking, upon previously paying a just indemnity.

There are excepted from this servitude the buildings, pleasure grounds, and gardens belonging to dwellings.

Article 2. The riparian owner of the lands upon which the right will have been claimed can always demand the common usage of the dam by contributing one half of the expenses of the establishment and maintenance of it.

Any indemnity will not be due in this case, and if any has been paid it must be returned.

When this common usage will only be claimed after the commencement, or the completion of the works, the payment which the second proprietor will have to make in order to have the right to use it, will be only that amount which it is necessary to expend in order to make it available for taking out water on his bank.

Article 3. The questions to which the application of the two above articles will give rise will be taken before the courts.

They will be proceeded with in a summary manner, and if there is need of experts, the tribunals will name only a single expert.

Article 4. There will be nothing detracted by the present provisions from the laws which regulate the police of waters.

APPLICATION OF THESE LAWS.§

The law of 1845 concerning the servitude of right of way to conduct water, and the law of 1847 concerning the servitude of right to construct a dam, were intended for application only in cases of individual or private works proposed, and unless their application is specially extended by law they cannot be availed of by companies or associations of land owners.

Two individuals cannot jointly force the application of these laws, though each for himself can. An association of landholders cannot avail themselves of these laws unless they organize according to the terms of a law of 1865, regulating the formation of syndicate associations, which expressly extends to such associations when duly recognized by the administration, the benefits of the laws in question. Hence "free" syndicate associations cannot force a right of way or a dam right, but "authorized" associations can.

The decrees of authorization of syndicate associations and the laws or decrees sanctioning the formation of canal companies, and grant-

---
* *Les Annales des Ponts et Chaussées*, Laws and Decrees, 1847.
§ See, De Passy, pp. 50, 89, 90, 100, 287, 314, and elsewhere.

ing them concessions of water privileges, always contain a clause extending to them the right, not only of eminent domain under the laws of 1833 and of 1841, to condemn lands for rights of way, but also the rights of laying the servitudes of right of way and right to abut a dam under the laws of 1845 and 1847, and it is usual to stipulate that lands for all main works shall be expropriated and paid for by them, and that only the servitude of right of way shall be acquired for minor works.

The right of way law cannot be applied to force an upper ditch owner to enlarge or deepen his existing canal in such manner as to pass sufficient water for other irrigations below; but it may be used to force any number of ditches through one piece of property, if the courts choose to allow its application for the purpose.

AUTHORITIES FOR CHAPTER VI.

In preparing this chapter I have consulted and compared the following named authorities:

Dumont.—[Work cited as an authority for Chapter II (French).] See Book II, Chapter V.

De Passy.—[Work cited as an authority for Chapter II (French).] See pp. 50, 89–100, 287, 314, and elsewhere.

Les Annales des Ponts et Chaussées.—[Work cited as an authority for Chapter II (French).] See Vol. XII, p. 328 *et seq.*, Vol. XX, pp. 203–217; also, Vols. Laws and Decrees for 1845 and 1847.

CHAPTER VII.—FRANCE[6];

IRRIGATION ENTERPRISE AND ORGANIZATION.

SECTION I.—*Governing Influences.*
　　Diversity of Climates.
　　Sentiment Concerning Irrigation.
　　Small Land-holdings.
　　The Agriculturists not Capitalists.
　　Jealousy of Property Rights.
　　Timidity in Regard to Indebtedness.
　　Heavy Cost of Works.
　　Poverty of Peasant Proprietors.
　　High Valuation of Lands.
　　Riparian Rights and Other Complications.

SECTION II.—*Irrigation Organizations.*
　　Speculative Companies.
　　Associations of Landholders.
　　Free Syndicate Associations.
　　Authorized Syndicate Associations.
　　Powers of Prefects and Principles of Association.

SECTION I.

GOVERNING INFLUENCES.

CLIMATIC AND SOCIAL.*

France lies in the zone intermediate between those latitudes, in Europe, where, on the one hand, irrigation is, as a general thing, an absolute necessity to success in agriculture, and where, on the other hand, it is useful only as an auxiliary to special cultivations, in limited localities and for particular purposes.

The climate of France, as affecting irrigation, is almost as varied as that of California; so that there are regions where the annual rainfall scarcely exceeds a foot in depth, and where it is so distributed, as to time, that there must be artificial waterings of all crops, to supply the deficiency of moisture to the soil and plant, and irrigation is practiced during the spring and summer months for this purpose.

And, again, there are regions, by comparison, quite cold, with twice

---
* See, Reclus, chapters "France"; also, Mangon, and De Buffon, Book I, Sec. I.

to three times as much rainfall as in those first spoken of, and distributed well throughout the year, but where irrigation is practiced far more copiously, and every month in the year, not to supply any deficiency in moisture to the soil and plant, but to serve as a fertilizer and as an equalizer of temperature to the grass meadows upon which extended dairy farm interests depend.

As a general thing, however, France is less an irrigation country from necessity and for general profit, than is California, for the valleys of France, with exceptions limited to small regions, receive from sixteen to thirty-two inches of rain each year, while ours of California receive only ten to eighteen inches, as a general rule.

The necessity for and value of irrigation was not sufficiently appreciated by the generations past, to bring about a general sentiment in favor of national encouragement to irrigation enterprise. Irrigation has been in France, as in California, until within comparatively few years, looked upon more as a local necessity, for some parts of the country, than as a valuable auxiliary to general agriculture, and as a process essential to higher and fuller agricultural development for all parts of the country. Hence, there has not been that widespread appreciation of the subject among the people of all France which we, not realizing these points, might expect to find recorded.

SMALL LAND-HOLDINGS AND JEALOUSY OF RIGHTS.[*]

The lands are very generally held in small tracts; and close and thorough tillage has taken the place of that wasteful, but easy, use of water, which is substituted for skill and industry in some other countries which might be mentioned.[§]

The generally humble condition of the peasant land proprietors, of south France particularly, and the minute subdivision of land, may be judged from the fact that when the association for the canal de l'Isle, department of Vaucluse, was set on foot in 1845, there were 1,414 subscribers, of whom 1,095 desired irrigation for tracts less than one hectare (2.47 acres) each, and 205 others for tracts less than two hectares, and, out of the whole number, only four subscribed for areas greater than ten hectares (24.7 acres) each.

The St. Julian canal, eighteen miles in length, irrigating from 6,000 to 7,000 acres of land, is the property of an association of irrigators, having 2,060 members; and the Crillon canal, irrigating 1,600 to 2,000 acres, has 750 subscribers to its construction and maintenance; these cases showing from three to three and a half acres in

---

[*] See, Moncrieff, pp. 38, 39, 61–63, 76, 77, Chap. II; also, Barral.

[§] It is not to be understood from this, however, that the use of water in France is particularly economical. As will be shown in a later part of this report, such is not the case.

one instance and from two to two and a half acres in the other as an average to the subscribing proprietor or irrigator.

"This minute subdivision of land seems to be at once the promoter and the hindrance to the extension of irrigation in France. It is these peasant proprietors alone, who till their own fields with their own hands, who fully appreciate irrigation." Without it their lands require less labor than can be put on them to advantage with it; and their spare time must be spent in labor for hire which is uncertain and not very remunerative.

With irrigation their time may be fully occupied on their own lands and their labors be rewarded by sure and abundant harvests.

The large land proprietor, on the other hand, who lets his land out to tenants, reaps less direct benefit from irrigation, for the tenants, alleging that much labor is bestowed on works that remain with the estate, refuse to pay materially higher rents by reason of irrigation facilities.

The greater appreciation of and desire for irrigation, by small proprietors than by large, is attested by the figures heretofore given for the case of the canal de l'Isle, and by the fact that in this case the small proprietors generally subscribed for water for the whole or at least half the areas of their lands, while the few large proprietors who interested themselves at all in the undertaking, subscribed for very small portions of their estates.

The larger landholders cultivate their fields in cereals and other crops not requiring irrigation, and taking less constant and skilled attendance and labor than do those irrigated; and, hence, as a general thing, in this south of France, where irrigation is most necessary, were it not for the desire of the smaller proprietors for irrigation on their tracts, many existing canals would not have been built when they were, or perhaps not at all.

HEAVY COST OF WORKS—POVERTY OF PEASANTS.[*]

And now, where irrigation has not yet been introduced, these peasant proprietors are poor and have no credit, individually; so that the want of capital among them, and the apathy of the larger proprietors, forms the greatest drawback to the further extension of irrigation.

In this condition of affairs a great trouble met with in the promotion of irrigation enterprise is the difficulty of securing subscriptions for water for a reasonably large proportion of any compact district, so that the lands subscribed for, being in small parcels and scattered,

---

[*] See, Moncrieff, Chap. II.

the works are made very much more costly to the unit of area irrigated than they otherwise would be, and the cost of maintenance and administration is greatly increased.

In the case of the canal de l'Isle, already spoken of, the total cost of construction for all works was estimated at about $23 per acre for lands subscribed for, as against $6 50 per acre if all the irrigable lands in the district had been subscribed for and the works made adequate to supply water for them.

HIGH VALUATIONS OF LANDS.*

Another great drawback to the advance of irrigation is the high price that land commands without water, and the high price of rights of way.

In the region spoken of, dry valley lands range in price from $300 to $800 per acre, while if commanded by a canal for irrigation, and having a subscription for water, they are worth only about thirty to fifty per cent more, according to circumstances.

Now, in California lands purchasable at $3 to $10 without opportunity or reasonable hope of irrigation, command $50 to $200 per acre when water is brought to them and they have the privilege at hand to receive and pay for irrigation.

There has been no such opportunity to speculate in lands in France, in connection with irrigation enterprise, as there has been in California, and, thus, a great incentive to the construction of works has not been present there that has been afforded here.

THE RIPARIAN RIGHTS QUESTION.

The riparian rights question which has come up, as we have seen, in a peculiar form in France, and the right of way question, also distinctive in its character, have held back irrigation enterprise immeasurably, but the conservative business temper and poverty of a large element of the agricultural population, and the indifference of the landed capitalists to the development of an industry which was calculated to render the care of estates more burdensome, has done much more to prevent advancement in this line of enterprise.

It has been the object and apparently the earnest desire of the government, not only to provide by legislation some means of directly meeting and setting aside the circumstances and retarding influences spoken of, but to impart an active impulse to agricultural development by enterprise in irrigation.

---

* See, Moncrieff, Chap. II; also, Barral.

It now remains to be seen what means have been employed with this view.

SECTION II.

IRRIGATION COMPANIES AND ASSOCIATIONS.

SPECULATIVE COMPANIES.*

Although not an invariable rule, the form of irrigation enterprise in France, and of government encouragement thereto, has been largely governed by the character of the stream—whether floatable or non-floatable—from which it was necessary to derive the supply of water in each case.

From floatable streams—dependencies on the public domain—the government, exercising the full right of state ownership, could authorize diversions by and encourage the construction of works on the part of any worthy applicant for concessions. And, hence, capitalized companies of non-landholders have sought and obtained sanctions and privileges for the construction of works from such streams.

The character of these organizations and their method of operation in the enterprises undertaken, will be of necessity sufficiently illustrated in the next section of this chapter, in speaking of the policy pursued by the government towards them, and, hence, nothing further will be said of them here.

ASSOCIATIONS OF LAND OWNERS.§

On streams not of the public domain another form of organization for works has been necessary.

Remembering that water rights for purposes of speculative canal enterprise, are not to be acquired on streams not declared navigable or floatable, that the waters are held for the bank lands, and that land holdings are, as a very general rule, in small parcels, we see that individual enterprise in canal building from such streams is kept within very narrow limits.

The waters are dedicated to the use of the riparian proprietors for the irrigation of their river lands—the water, in a measure, is attached to the lands, and cannot be alienated.

A proprietor by buying back land next adjacent to his bank land, can to some extent increase the width of his irrigable area, but the courts and the administration—the one restricting the extent of his

---

* See, De Passy, pp. 103-130; also, Dumont.
§ See, Dumont, Book II, Chap. VI, Sec. I; De Buffon, Vol. 2, pp. 89-98; De Passy, pp. 79-102.

water privilege, and the other the size of his headworks—would very soon stop any attempt at an extension in this way which was not equitable to other proprietors.

Furthermore, rivers of this class in France generally run in valleys whose lands slope down towards the streams (and not, as do many streams in California, across plains which slope back each way from the stream), and, consequently, canals of short length cannot command any considerable width of territory for irrigation.

These circumstances have resulted in the construction of a great number of very small ditches, where, as is frequently the case, the grade of the streams has been sufficiently rapid to admit of the water being brought out upon the land within the limits of one, or at most, several land holdings.

The scope of these individual and partnership enterprises has been, until within a few years in the past, still further restricted by the absence of any legal means of acquiring right of way for a canal through, or right to build a dam on or next to the lands of others.

The leading writers on irrigation dwell upon the great drawback to irrigation in France, which has resulted from these circumstances.

Furthermore, the simple partnership association which would answer as a business arrangement between several neighbors, for the construction of a little private ditch, would not answer for the organization of a large enterprise for the benefit of perhaps several hundred or thousand land holdings.

The French agriculturists appear to have been extremely jealous and careful of their rights; desiring to have and hold them, as near as possible, immediately under their personal control, and hence have not adopted forms of association which would be popular in this country.

These circumstances led to the passage of laws recognizing the form of organization known as a *syndicate association*, which is that now generally adopted by landholders for the conduct of works on joint account, necessary in the development, in any way, of agricultural neighborhoods.

A syndicate association is a society of land owners, organized according to general forms prescribed by laws and decrees, but with terms of organization arranged according to the will of the members, as embodied in the articles of association.

AN ANALYSIS OF THE LAW OF ASSOCIATION.*

The law recognizes eight purposes for which syndicate societies may be formed, as follows:

First—The construction and management of embankments and other works for protection against the sea, torrents, and the waters of non-navigable rivers.

Second—The cleansing, deepening, straightening, or regulating canals and water-courses not navigable nor floatable, and of irrigation and drainage canals.

Third—The construction and management of works for the drainage of fresh water marshes.

Fourth—The construction and management of works for the reclamation of salt marsh lands.

Fifth—The construction and maintenance of works for the sanitary improvement of wet and unhealthful districts.

Sixth—The construction and management of works for irrigation and *colmatage*.

Seventh—The construction and maintenance of works of land drainage.

Eighth—The construction, maintenance, and management of roads and every other improvement of agricultural lands and neighborhoods, which requires coöperation amongst proprietors.

The general organization of associations is the same for all of the purposes specified, but the details of agreement and administration differ with the object in view. The forms and provisions ordinarily followed and adopted in and by associations for irrigation, only, will be spoken of here.

The law recognizes two kinds of syndicate associations: The first called "free," because held together only by the expressed will of the members; and the second called "authorized," because specially declared, in each case, to constitute an organization of public utility, and so "authorized" to exercise the right of eminent domain in condemning private property for the purposes of the association.

These societies are formed upon the basis of the land to be beneficially affected by the works contemplated; representation and voting power in the general assembly of subscribers being proportioned somewhat to the area held, varying in different cases, within prescribed bounds, according to circumstances and as determined and settled in the constitution or articles of agreement of the society.

Their boards of directors called *syndics*, constitute the *syndicate*

---

* See, particularly, De Buffon and De Passy, as cited; also, law of June 1, 1865, Decree of November 17, 1865, and the Ministerial Regulation—Appendices 2, 3 and 4, De Passy.

proper, although the whole association is frequently called a syndicate. Being legally constituted bodies, they can enter into court, acquire or dispose of, exchange or hypothecate property, and do all that an individual might do in a business way.

FREE SYNDICATE ASSOCIATIONS.

Free syndicate associations are formed by the declaration of the associates, and the signing of the agreement of association, etc., as follows:

The agreement or act of association specifies the object of the enterprise, regulates the mode of administration of the society, and fixes the limit of authority confided to the administrators or syndics. It determines the ways and means necessary for the raising of funds, and the mode of collecting assessments or subscriptions.

It must be published in a journal of official announcements, and copied into the records of the prefecture.

In the case of an association formed for the construction, maintenance, and management of irrigation works, all proprietors of lands susceptible of irrigation, within the district, must be admitted as members should they desire to join; each designating the lands and the area thereof for which he desires to subscribe.

The volume of water conceded is ordinarily divided amongst the proprietors in proportion to the area subscribed for, and without reference to the kind of crop or character of land cultivated and worked. These terms being fixed by the articles of association in each case, and not by the law, are variable, according to the will of the associates.

The right of irrigation goes with the land subscribed for, and cannot be alienated or passed to other lands.

Each associate is bound to accord right of way for ditches through his land, upon payment of indemnity fixed by arbitration. Thus, law suits are avoided on this score.

Each associate is a member of the general assembly, having voting power according to the terms of the agreement in each case. Sometimes the vote is by units of land area between certain limits, a minimum area and a maximum area to a vote, or, for instance, one vote to each holder of from one to five hectares. Thus the proprietor owning between one and five hectares, would have one vote each; those between five and ten hectares, two votes each, and so on, a vote to each five hectares or fraction not less than one hectare.

The general assembly elects directors, called *syndics*—five, seven, nine, or more, as the case may be—who form the *syndicate*, or board of management of the association.

In some organizations the syndicate is all powerful—in others, many questions have to be submitted to the general assembly for final settlement.

The syndicate name from their number a manager or general director, who is the chief executive officer of the association. — —

Other officers, as secretary, treasurer, etc., are similarly named, as in societies whose organization is familiar to everybody.

The syndicate employs an engineer, and all projects for works are duly and completely drawn up and adopted by the board before construction is authorized.

The cost of works and expenses of management are ordinarily borne in proportion to area subscribed for, and without reference to value of lands or crops, or character of cultivation or soil.

Assessments under the law, are made collectable as taxes, and are a lien on the property subscribed for.

AUTHORIZED SYNDICATE ASSOCIATIONS.

All syndicate associations must be first formed as free associations, and they may then apply to the administration for recognition as authorized associations.

The prefects of the provinces have authority to make these decrees of recognition and authorization, following after certain forms and instructions embodied in decrees and laws of the general government.

The application to the prefect must be accompanied by plats of the proposed district, including the lands to be irrigated, each parcel being designated and tinted with a color representing its condition as to cultivation, soil, etc., and whether or not it is irrigable, and if so whether or not it is subscribed for in the association.

A list of subscribers accompanies these plats, and a statement of the financial ability of the subscribers to meet their engagements.

A regular project for works and for financial management is also submitted, from which to judge of the feasibility and cost of the scheme and the adequacy of the organization to carry them out.

The law provides that the desire of the members of the free association, to have it converted into an authorized association, must be expressed in general assembly, as follows: "If the majority of the individuals interested, and representing two thirds of the area of land subscribed for, or if two thirds of the individuals owning more than one half of the area of land subscribed for," desire the change, the prefect, being satisfied of the soundness of the enterprise in other respects, issues the decree of authorization.

The application must show, in addition to all the above, the plan of the organization, the plan of representation in the general assembly and the basis for voting, as well as the basis for the division of expenses.

Following this application a public announcement is made. The application is published and the plans, etc., are opened to inspection and comment and everything opened to objection.

Each proprietor of lands affected is notified as to the application, and requested to appear at the prefectorate if he has any objections or criticisms to make.

A register is exposed, in which every interested party may write his remarks and criticisms.

A commission of landholders not interested is appointed to report on the results of the examination.

These and other formalities, taking a month or more according to circumstances, being gone through with, the prefect considers the case and renders his decree of authorization or refusal.

The action of the prefect one way or the other, is appealable from to the minister of public works.

PREFECTORIAL POWER—GOVERNMENT POLICY.

Prefects may refuse to issue decrees of authorization for associations, for various causes, amongst which are the following:

The district not being large enough to render its works of public utility.

The works proposed themselves not being sufficiently important to justify the foundation of an authorized association.

The district not comprehending the area it should take in, and other proprietors desiring to come in.

The lands within the district not being sufficiently subscribed for.

In the case of authorized associations the government in a measure becomes accountable for the meeting of their engagements, so that the assessments are not only collectable as taxes by the officers of the syndicate, but the government authorities, if necessary, may interfere and force their collection so as to make good the debts of the district.

Condemnation of lands for the benefit and use of the association is conducted by the syndicate in conformity to a general law providing for the condemnation of private interests for the public good, but this can be done only after a declaration of public utility has been made in favor of the proposed works in each case by the council of state.

In cases where the association asks a subsidy from the government funds, or from those of the department, it is always provided that the prefect may name a number of syndics to represent the state or the department in the syndicate, in proportion to the part of the whole cost of the works which the subsidy provides for.

In cases where the association is formed for irrigation, or any purpose where water is desired as an auxiliary to some operation to be carried forward, the formation of the society may be had for only a portion of the district embraced within the exterior limits of lands subscribed for, but in cases where, as in reclamation or drainage, all of the lands in the district are necessarily affected by the works, the whole area is brought under contribution, and when two thirds of the land is subscribed for, the other third is forced to contribute its share to the expense.

This rule is the outcome of a long struggle in France, in which it has been proven, that some landholders will always hold back and prevent necessary public improvements, and that the interests of the public demand, in cases of reclamation and drainage, that they be made to join in with the majority in their district, or sell out to those who will carry forward the works. And the tendency of events and sentiment is towards a similar policy with respect to irrigation districts, also.

AUTHORITIES FOR CHAPTER VII.

In the preparation of this chapter I have consulted and compared the following named authorities:

Reclus.—[Work cited as authority for Chapter II.] See chapters, "France."

De Passy.—[Work cited as an authority for Chapter II.] See, particularly, pp. 79–102, 103–130, and appendices 2, 3, and 4.

Dumont.—[Work cited as an authority for Chapter II.] See, particularly, Book II, Chap. VI, Sec. I.

De Buffon.—[Work cited as an authority for Chapter II.] See Vol. I, Sec. I; Vol. II, pp. 89–98.

Barral.—"Irrigation in the Department of the Mouths of the Rhone." By J. A. Barral; being an official report of a Government Commission of Inquiry into the subject of the use of Waters in Irrigation in France; 2 vols. quarto; Paris, 1876–77.

Barral.—"Irrigation in the Department of Vaucluse." Same set of reports as the preceding; 2 vols. quarto; Paris, 1877–78.

Mangon.—"The Employment of Water in Irrigation." M. Hervé Mangon, a Chief Engineer in the Government Corps of Civil Engineers, France; 1 vol.; Paris, 1869.

Moncrieff.—"Irrigation in Southern Europe." By Lt. C. Scott Moncrieff, Royal Engineers, Great Britain; 1 vol.; 8 vo.; London, 1868. See, particularly, pp. 38, 39, 61–63, 76, 77.

CHAPTER VIII.—FRANCE[7];

GOVERNMENTAL POLICY AND IRRIGATION CONCESSIONS.

SECTION I.—*Features of Policy and Forms of Enterprise.*
    Political and Social Conditions.
    Forms of Governmental Encouragement of Irrigation.
    Early Irrigation Enterprise.
    Tax Rebate on Advanced Values.
    Subsidies, Advances, Loans, and Guarantees.
    Prize Competition in Irrigation Practice.
    Statistical Atlas of Irrigation.

SECTION II.—*Notable Instances of Enterprise and Encouragement.*
    The *Canals*—Des Alpines, Carpentras,
    Cadenet, St. Marterey, Siagne,
    Siagnole, Bourne, Rhone,
    Vesubie, Pierre-latte, Manosque,
    Herault, Ventavon, Petite-Vence,
    Malpas, St. Marcel, Argeliers, and Raouel.

SECTION I.

FEATURES OF POLICY AND FORMS OF ENTERPRISE

POLITICAL AND SOCIAL CONDITIONS.

The French government, although apparently always appreciating the value of irrigation to all France, and directly favoring irrigation enterprise, as we shall see, by several important measures of policy, has not, as in the case of interior navigation and the promotion of arterial drainage and consequent land drainage or reclamation of lands, directly taken the lead in the construction of works for the purpose, at public expense and wholly under national management, except in cases where the submersion of vines to exterminate the phylloxera vine pest was a ruling consideration, or in districts where the landholders were exceptionally poor and without credit.

Rivers were improved and made navigable where before unfitted for the purpose, and great canals constructed for navigation, as public works of the nation, more than a century ago. The policy which prompted this action has ever been in the ascendancy, and was quite fully developed under the last empire, and has been renewed and enlarged upon by the present republic; but towards irrigation, the policy has been rather to encourage the efforts of landed proprietors in constructing their own works, or to encourage the investment of capital in irrigation enterprises upon terms such that, at the expiration of long periods of years, the works should revert to associations of the owners of the lands irrigated, or to the central, departmental, or municipal governments, for the benefit of the people.

It is to be remembered that all of the irrigable lands were in private ownership—the government not having any irrigable public domain—and that in the view of men of broad ideas, such as the rulers of the country have probably been, all France was an irrigation country, and should the government undertake the construction of works for the irrigation of one section, without some specially potent reason, it should for equally good reasons bring water to the irrigable lands of all the people.

Furthermore, the French agriculturists, although largely composed of a peasantry inferior to American farmers in enterprise, wedded to old habits and customs, and comparatively slow to take up with and realize the lessons of experiences had elsewhere, have never stood in that relation to their government, which those of Egypt and of India, where nearly all irrigation works are built and managed by the governments, have to theirs. The French government realized this difference in the people and the political and social conditions of countries, when in its province of Algeria it pursued a different course towards irrigation, and, following in the footsteps of the khedives of Egypt and the English rulers of India, constructed great irrigation reservoirs, canals, and ditches, as public works of the nation.

FORMS OF GOVERNMENTAL ENCOURAGEMENT.*

As we have seen, irrigation enterprise has taken three forms:

First—In the construction of works on private account for the benefit of private lands, by one or several land proprietors jointly.

Second—In the construction of works for the common good of the owners, by associations of land proprietors.

Third—In the construction of works by individuals, companies, or municipalities, for the distribution and sale of water to consumers.

The government has encouraged all these forms of enterprise, and has also encouraged the skillful and economical use of water in irrigation by the individual irrigator.

---

* See, De Passy, Dumont, and Malapert; but, particularly, the various laws making concessions to companies, societies, and associations, as hereafter quoted, and many others to be found in the volumes of the *Annales des Ponts et Chaussées*.

This policy of direct encouragement has in application taken various forms, as follows:

First—A remission of tax assessments, for certain long series of years, on the increase of land valuations due to irrigation.

Second—The loaning of funds on most favorable terms to companies or associations undertaking irrigation works.

Third—Advancing to such companies or associations a large part of the cost of their works and taking the works themselves in payment at the expiration of long term concessions.

Fourth—Subsidizing enterprise in the construction and management of irrigation works, by payment of large sums to the sole benefit of the companies or associations, or that of the departments, municipalities, or irrigators ultimately acquiring ownership of the properties.

Fifth—Guaranteeing interest on capital invested in or borrowed on great irrigation works.

Sixth—Construction of main irrigation works at state expense and turning them over to syndicate associations for management.

Seventh—Construction and management of irrigation works wholly as public works of the state.

Eighth—The inviting of competition in and granting premiums for the best irrigation practice.

Ninth—The collection of irrigation statistics and useful data of irrigation practice and the publication thereof for general information.

EARLY IRRIGATION ENTERPRISE.*

The first works of irrigation, other than purely individual enterprise, constructed in France, were made under grants of right from the counts, and were combined with and secondary to those for water-power purposes.

Thus, in 1171 Raymond V, Count of Toulouse, in the south of France, granted to the bishop of Cavaillon the exclusive right to divert water from the Durance, a river carrying 3,000 cubic feet per second at its low stage, into canals for the purpose of supplying power for cornmills to be constructed. The bishop constructed a work known as the St. Julian canal, and sixty-four years afterwards granted to the inhabitants of Cavillon the right, for which they had applied, to use the waters in irrigation. This led to an enlargement and extension of the canal and an agreement as to the distribution of expense for maintenance of the work, and to this day, in accordance with this ancient usage, those who use water for power pay one third

---

*See, M. Conte in *Les Annales des Ponts et Chausseés*; also, Barral.

the annual expense, while those who use it in irrigation pay the other two thirds.

This was one of the first, if not the first enterprise of which there is record, in which a trace of encouragement to irrigation on a large scale is to be detected.

Permits for water for irrigation from canals constructed by government for purposes of navigation, were granted in the early period of public works enterprise, but these were for very small quantities of water and to individual farmers or small communities only.

At a later period in the construction of some government canals for navigation, irrigation, as well as the supply of water for motive power, for industrial uses, and municipal domestic purposes, was considered, and the works planned so as to produce a current from the main source of supply, such that while navigation was not impeded the other interests were to some extent subserved. But these instances have been exceptional in the planning of public works, and it cannot be said that irrigation has received generally any material help in this manner until within the past few years, when quite a number of small canals have been built out from the main canals of navigation for purposes of irrigation, but largely with the immediate view of preventing the spread of or destroying the phylloxera in the vineyards.

TAX REBATE ON ADVANCED LAND VALUES.

The first form that direct encouragement to irrigation enterprise on a large scale took in France, was that of an engagement, on the part of the government, not to raise the assessed valuation of the lands brought under irrigation above what it had been before irrigation, for a period varying from twenty to thirty-five years after the waters were introduced on them, and not to tax the works of irrigation at all for some such like period, but only to assess the lands occupied by them, as they had been assessed before.

In the case of the Carpentras canal, in south France, constructed in 1853–54, this period was fixed at twenty-five years after the construction was completed, according to plans, specifications, and agreement.

This was a measure of encouragement more especially in the interest of the land proprietors, who would unite in an association, under terms of law, over a sufficiently large area to promise a development calculated to be of future importance to the country, and who, under government supervision, undertook to construct substantial works to insure such development by the irrigation of their lands.

Even with this encouragement, irrigation made slow progress in France. Great areas of country stood much in need of it; other considerable regions were in a condition to be greatly benefited by it, but the spirit of enterprise did not seem to take hold of the landholders generally in the cause.

The rich did not want to adopt a system of agriculture calculated to make advisable the expenditure of much more labor on their farms, and to require a much closer attention to their estates; and they did not generally appreciate the moneyed value of irrigation properly conducted.

The poor landholders in many quarters were not awakened to the results of experiences in irrigation favorable to their class in other quarters, were wedded to old habits and customs, were jealous of the slightest move calculated in any way to interfere with their full control of their little home grounds. They did not understand and could not appreciate the benefits of association of interests for common good in districts. They each would like to have a canal or ditch of their own, but did not want to join with several hundred or thousand others to get one jointly.

Further than this, their poverty often, though their conservatism were overcome, stood in the way of their undertaking large works, even when they might combine for the purpose; and, as we have seen, the laws themselves hampered the spirit of enterprise on nonnavigable streams, by the water-right complications which their riparian right and other rules had brought about.

SUBSIDIES, ADVANCES, LOANS, AND GUARANTEES OF INTEREST OR INCOME.

Government encouragement then took the forms of loaning funds for long terms on irrigation works, advancing part of the cost of the works, and taking the works themselves in payment at the end of long terms, and subsidizing large irrigation enterprises, without return other than nominal.

These forms of encouragement were more directly intended to give irrigation projects good financial standing, and to enable capital to enter into the field of enterprise with a certainty of a moderately good return.

Being incident to the construction and management of works, and not to the ownership or tillage of lands, these measures addressed themselves to capitalized companies or societies; and a number of such organizations have sought and taken up with government offers of this kind, binding themselves to construct works according to prefixed and approved plans, to maintain and manage them under prearranged regulations and government supervision, to deliver water for irrigation, etc., at predetermined rates, and, finally, to return the money borrowed, or turn over the works to the government or a department, or, perhaps, to a syndicate of landholders, at the expiration of the term of the concession. Direct subsidies, without return, have only been granted to syndicates of landholders, and presumably in cases wherein their financial condition was poor and their credit bad, and not to capitalized companies, as in the cases of the encouragements by loans or advances on cost of works.

These measures of encouragement brought about also the organization of a number of associations of irrigators, who have sought to derive not only the advantages of the first measure, in the limitation of taxing valuations, but also the benefits of loans, advances, and subsidies.

PRIZE COMPETITION IN IRRIGATION PRACTICE.*

The final measure of financial encouragement to irrigation which the French government has instituted of late years, is that of giving premiums for the best examples of irrigation practice in the several great irrigation centers, the convoking of meetings of irrigators and land owners on the occasion of making the examinations of competing tracts, and the publication in great detail of all valuable and practical facts about irrigation acquired by these examinations, meetings, and discussions, as a supplementary act of interest.

From the first report of Mr. Barral, the reporter of the commission or jury appointed to conduct the first of these proceedings in the department of the Mouths of the Rhone in 1875, I take the following general account of the origin, purpose, and progress of the movement:

"The minister of agriculture vividly impressed with the role that irrigation plays in the practice of agriculture in the south of France, and the necessity for showing to the agricultural population all that can be derived from irrigations properly conducted, in the interest of individuals or the wealth of the country, resolved to institute for five years in the department of the Mouths of the Rhone, a convention of agriculturists, whether proprietors or renters, who have used waters from the different irrigation canals in an intelligent manner."—[Barral, Vol. I, Chap. I.

Prizes and medals were promised by a decree of June 2, 1874, to those whose use of irrigation waters could be shown to have been the most systematic, economical, effective, and remunerative.

This decree of the minister of agriculture, representing the government, was as follows:

"The minister of agriculture and commerce, with the object of encouraging the efforts that tend to the progress of agriculture, and especially to cultivation by irrigation, looking at the losses occasioned by phylloxera, and the necessity to transform or increase the produc-

---

* See, Barral, particularly, Chapters I and II of each volume.

tion of irrigable land; looking at the notice of the inspector-general of that region (the Mouths of the Rhone); on the proposition of the director of agriculture, issues this

"DECREE :

Article 1st. Rewards are offered in the department of the Bouches du Rhone, in 1875-76, '77, '78, and '79, to agriculturists, proprietors, or renters, who have utilized in the most intelligent manner the water of the different irrigation canals.

"Article 2d. These rewards are divided in the following manner:

"FIRST CLASS—Properties containing more than four hectares (about 10 acres) of irrigated land—

"1st prize—Gold medal, and 1,000 frcs. ($250).
"2d prize—Silver medal, large size, and 700 frcs. ($140).
"3d prize—Silver medal, and 600 frcs. ($125).

"SECOND CLASS—Properties irrigated to an extent of four hectares and less—

"1st prize—Gold medal, and 600 frcs. ($125).
"2d prize—Silver medal, and 500 frcs. ($100).
"3d prize—Bronze medal, and 300 frcs. ($60).

"Article 3d. A work of art would be bestowed on the winner of the first prize of one of the above classes, if recognized or judged worthy of being specially rendered noticeable for the economical management of water in the practice of irrigation. In case of the gift of the work of art, the gold medal, for first prize, will not be bestowed.

"Article 4th. The statement of the contestants, containing an explicit note and an exact indication of the extent irrigated, certified by the mayor of the commune, must be addressed to the prefecture of the Bouches du Rhone on March 1st, current year, at the latest.

"Article 5th. The director of agriculture is charged with the execution of the present decree.

"Made at Versailles, June 2d, 1874.

"L. GRIVART, Minister, etc."

In 1875, besides the very general interest awakened amongst all agriculturists in the region, there were thirty-nine competitors for the rewards or prizes, and each property and system was made the subject of special study by the commissioner or jury.

M. Barral says: "This study presented great interest. The question was not only that of the competition for the prizes, but was also that, which is of a higher order, of ascertaining the services rendered by the water in giving a more abundant production, and in the protection of vines against the attacks of the phylloxera.

"The examples of irrigation practice reported are of the highest importance to agriculture, and of great use to those who are in a position to usefully employ water in cultivation.

"The circumstances under which these cultivations were found, in response to the offer of the government, are varied enough to justify the drawing of general conclusions from the facts observed. These conclusions show that a great increase of wealth would be the result for national agriculture, were works undertaken on all water-courses capable of being transformed into irrigation canals or capable of feeding such canals.

"In view of these things the judge-advocate of the jury received orders to enter into all the details of the subject. His statement must contain all information needed in the practice of irrigation, and, also, all that might be of service to the officers of the public administrations.

"With the view of making a network of canals all through the country, it is important to encourage the forming of companies or associations having power to construct works, and to develop amongst the rural population the habit and skill of using water in irrigation systematically; to incite land owners to engage in irrigation enterprise and advance the funds necessary for the diversion of large streams with the certainty of receiving considerable profit from it.

"The object of this report, therefore, is not only to point out by the proof of facts that can easily be verified, the justice of the decisions in the competition of the year, but to make known to all agriculturists, and to land owners, what an enormous source of wealth water is, and in particular, that to-day it has become, in a great many places, the providential means of saving the vineyards from the attacks of an underground enemy (the phylloxera) which threatens to make them disappear."

To give an idea of the extent of and importance attached to this governmental move in the interests of irrigation, I mention the fact that the commission, or "jury," placed in charge of the examinations, awarding of premiums, and reporting results, was composed of (1) the inspector-general of agriculture of France, president; (2) a deputy inspector-general of agriculture, vice-president; (3) the life secretary of the central society of agriculture of France, reporter; (4) the general secretary of the society of agriculture of the department of the Mouths of the Rhone; (5) the director of the agricultural college of Paillerols, Lower Alps; (6) an engineer of the government civil engineer corps, and (7) the vice-president of the society of agriculture of the department of the Herault. And it is further notable in this connection that the reports of this commission for the three years of 1875–76–77 take up four large quarto volumes, containing 1980 pages of printed text and numerous maps and tables.

The decree quoted was for the one year of 1875, and the one department of the Mouths of the Rhone, and was not only followed by a similar decree and concourse and awarding of prizes each year in that department, but, also, by like action in other departments; as, for instance, in that of Vaucluse for which the first action was taken in 1876.

Thus, gradually progressing through all the departments where irrigation is practiced, the government, through its department of agriculture, is not only making this most thorough and intelligent study of the use of water in irrigation, but is directly encouraging the irrigators, in the bringing of them together for discussion, by awarding prizes for the best examples of irrigation practice, and by publishing in detail all the data thus acquired.

And further than this, not stopping at an examination and study of practice at home, this department has sent well trained and intelligent agricultural engineers to other countries where considerable progress is being made in the use of water in irrigation, with instructions to personally study the systems and the practice, and collect all available data, in print and by verbal communication, that may be worthy of attention in the endeavor to enlighten and encourage its own agriculturists and guide its legislative and administrative officers.

Thus, the irrigation works and practice of California, in common with those of others of the United States, have been recently inspected and studied by a special agent, and, in common with other points where information might be had, the office of the State Engineer has been quite thoroughly examined and data collected therein; and all for the benefit of the irrigators and the agriculture of France.

STATISTICAL ATLAS OF IRRIGATION.*

And still again in another channel, we find the spirit of enterprise and enlightenment moving the French government in this connection. Under a ministerial order issued in 1869 a special commission, composed of nine civil engineers and scientific and practical agriculturists of high standing, was appointed for the purpose of "revising, coördinating, and preparing for publication the statistics relative to the amount of water available in the streams, and the use made of it in the various departments of France."

It was ordered that the chief of engineers should instruct all departmental engineers and conductors engaged in the hydraulic service to collect and forward the information desired from their several fields of operation, according to certain prescribed forms, and that this data should be turned over to the commission for its work.

The investigation has been progressing continuously, but is yet unfinished. Several partial and local tables have been published, and a set of eighty-five departmental hydrographic maps, which form the basis for the study, have been issued for the use of the collectors. The work is formulated with the view of treating the regulation of the waters and their use, as a business proposition. The government undertakes to find out exactly what waters are available from year to year, and exactly what is done with them. The work once done can be kept posted from year to year with comparatively light work and expense, and will furnish that data from which economy and efficiency can be studied and published. So that if there is

---

* See, Ministerial circular, July 4, 1878, *Les Annales des Ponts et Chaussées*, Vol. CXXXIX, p. 1122.

water available it will be publicly known; if there is water wasted and used unskillfully, it will be publicly known in a way to rebuke the users; if there is water used with economy and skill worthy of special note, it will be publicly known in a way to reflect credit upon those who thus utilize it.

There can be no question but that this is the real way to regulate the use of waters. Public knowledge of what is good in practice will bring imitation and economy as an average outcome; public knowledge of what is reprehensible and wasteful will bring condemnation, and a reform of the wrong.

These are, in substance, the sentiments to be found in late French state papers relating to irrigation, and with the expression of them, I leave the subject of the progress of French governmental policy towards the irrigation interest, for they are the evidence of the crowning feature of a long line of intelligent actions of a government fully awake to the best interests of its people.

SECTION II.

NOTABLE INSTANCES OF ENTERPRISE AND ENCOURAGEMENT.

In this section I present a series of abstracts of the laws authorizing, and the decrees and agreements regulating the construction and maintenance of the most notable canals of irrigation in France. It will be seen that they are scattered in date over the period of the past fifty years, and in character range the whole field indicated in the preceding section.

The study of these measures, together with that connected with the canal of the Bourne, of which a closer and systematic abstract has already been given, will lead to an appreciation of the fact that irrigation is a subject for careful and thoughtful treatment at the hands of the legislator.

THE CANAL DES ALPINES.*

In the year 1839 the concession for the northern branch of the Alpines canal and its secondary ditches was offered for sale. The concession was perpetual, and allowed five cubic metres of water per second to be derived from the river Durance, in time of ordinary low-water, in addition to the right formerly authorized on the portion of the said branch already opened. The concessionary was authorized to receive as his profit a rent from the irrigators which should not

---

* See, Royal ordinance of July 9, 1839. *Les Annales des Ponts et Chaussées*, Vol. XVII, p. 289, *et seq.*

exceed a litre and a half of corn of the country of best quality for each *are* (0.025 acres) of land irrigated, regardless of its nature.

The buyer could expropriate lands for the construction of the canal and its branches, in accordance with the law; and the owners of lands to be irrigated by the waters of the canal were freed from any increase of landed taxes over that then paid on them, for twenty-five years from the time fixed for the completion of the canal.

General plans had to be presented within one year from time of sale; the works to be commenced within six months from the governmental approbation of the project, and to be executed within six years from the final consummation of the sale. Forfeiture to be incurred for failure to comply with either or both of the two last mentioned conditions.

The landed tax was established on the canal for only the actual ground occupied by it, rated as lands of the first quality. The portion of water conceded which in the space of twelve years would not have been employed in irrigation, was to return to the disposition of the State, which could make it the object of a new concession.

The buyers were obliged to deposit, after the sale, in the treasury, the sum of 50,000 francs ($10,000). This sum to be increased to 100,000 francs ($20,000) in the three months which follow the approval of the sale. The said sum to be returned in fourths, in proportion to the amounts of work executed, and in case of forfeiture, the portions of the security not returned, to be confiscated by the treasury.

The enterprise was sold on the twentieth of June, 1839, to three individuals, as agents for the "General Drainage Company," with an abatement of two thirds per cent; that is, the annual rent was to be one litre forty-nine centilitres of corn for each *are* (equivalent to 54 quarts per acre) of land irrigated.

CANAL OF CARPENTRAS.*

The government was authorized to concede six cubic metres (212 cub. feet) of water per second to be taken from the river Durance and used in the irrigation of lands belonging to the communities of Saumannes, l'Isle, and others. The water could, however, be cut off from the canal by order of the prefect whenever such measure was deemed necessary either for the interest of navigation or for the protection of the interests of those who had previous claims to the water.

The enterprise was declared of public utility and the canal only taxed for the actual ground occupied by it, classified as of first quality. The lands to be irrigated from the canal were not to have their taxes

---

* See, Royal order of July 9, 1852, *Les Annales des Ponts et Chaussées*, Vol. XLVIII, p. 523, et seq.

raised over the assessment at that time, for twenty-five years from the date of the completion of the canal.

CANAL OF CADENET.*

This concession was made to a number of irrigating proprietors forming a syndical association, and consisted in permission to derive three cubic metres of water from the river Durance, and authority to contract loans to be first approved of by government or by the prefect, provided the debt of the syndicate did not exceed 50,000 francs ($10,000) at the time the loan was asked for.

This syndical association was called the "Society of the canal of Cadenet," with the object of irrigating certain lands belonging to the subscribers thereto, but as there were many persons and communities whose lands could be irrigated by this canal, but who did not subscribe, this decree provided that these parties could join the society either during or after its construction on the same terms as the original founders.

The society was administered by a syndicate composed of seven members to be named by the prefect. One of the members was also named by the prefect to fill the place of director of this syndicate, and attend to the business in connection with the construction, maintenance, and operation of the canal.

The enterprise was declared of public utility, but did not receive any assistance from the government, either in the shape of subsidy or remission of taxes. On the contrary, it seems to have been burdened with conditions of which the following are the most important:

1. Four tenths of a cubic metre of water, per second, had to be returned into the Durance by the escape canal of Pertuis.

2. The waters of the canal not utilized for irrigation had to be returned into the Durance at a specified point thereon.

3. It had to carry out all its works in conformity with the direction of engineers appointed by government, but paid by the society itself.

THE CANAL OF ST. MARTERY.§

The canal of Saint Martery, under a law, agreement, and schedule, passed and ratified in 1866, was conceded to three individuals, representing a company of English capitalists, called the *General irrigation and water supply company of France,* for a period of fifty years, and thereafter to belong in perpetuity to the department of the Upper Garonne, wherein it is situated.

---

* See, Royal decree of November 18, 1854, *Les Annales des Ponts et Chaussées,* Vol. LVI, p. 179.
§ See, *Les Annales des Ponts et Chaussées,* Vol. LXXXVIII, p. 162, *et seq.*: Law of May 16, 1866.

The canal and all its secondary and distributing works were to be built by the company, at its sole expense and risk, and managed and maintained by it during the term of the concession, and thereafter by the department.

The general government granted a subsidy of 3,000,000 francs to the work, to be paid in tenth parts, in proportion to the advancement of the principal canal, but depending on the resources available to the administration from time to time for such purposes. The payments were to be made on the certificates of the engineers, to the effect that a greater sum had been expended on the works, etc., since the last payment than the amount of the installment demanded. A reserve of 500,000 francs was to be made from the two last installments, of which 300,000 francs were to be paid over upon the final approval of the main canal, and 200,000 after the final approval of the secondary canals, which final approvals were to be one year after the claim of completion and the provisional reception of the works.

Complete final plans for the main work were to be submitted to the administration for approval, and the works to be commenced, under pain of forfeiture of rights and guarantees, within one year of the date of the concession; the main canal to be finished within five years, and the secondary canals within two years after approval of locations; and the company was required to deposit the sum of 150,000 francs as security for the faithful performance of its engagement.

To assist the company's credit for the securing of capital necessary for the construction of the works and the other purposes of the enterprise, the department of the Upper Garonne was authorized to and engaged to contract, upon the demand of the company and for its benefit, with the Credit Foncier of France, under a law authorizing such negotiations, one or more loans to the maximum amount of 4,000,000 francs. These loans were to be contracted upon the basis of the company's assured income from subscribed water rents, after the works were completed to deliver the water, and the collecting and management of the income was to be assigned to the department for that purpose, provision being made for maintenance and operation of works, the company at the same time pledging its faith and credit in the protection of the department from loss or embarrassment on account of the loans. The estimated irrigable area was 14,000 hectares—about 34,600 acres.

Subscriptions for the use of water were required to be made for fifty-year periods; the right of irrigation belonging to the land subscribed for, and going with it, no matter into whose hands it passes, not being transferable to other lands by the owner, and not forfeitable by the original lands after the expiration of the fifty years except by the owner's consent.

The quantity of water to be furnished for irrigation was fixed at three fourths of a litre per second per hectare (equivalent to a duty of 93.25 acres per cubic foot per second); and provision was made for a rebate on the rents in case of an insufficiency of supply for any term of more than thirty days duration during the six months of the irrigation season.

The price of water for irrigation was fixed at 25 francs per hectare ($1 92 per acre) per year, for all subscriptions made during the first two months of the examination of the project, at 35 francs ($2 69 per acre) for subscriptions made after that time and before the promulgation of the schedule, and 50 francs ($3 84 per acre) for all subscriptions made thereafter.

It was stipulated that sale by the quantity of water might be substituted for sale by the surface of land irrigated, and the substitution was to be made of the prices named above, for the *half* litre of water per second, with the provisions that water thus taken should be used only on contiguous tracts, that all of the lands should be pledged for the payment of the water rent, and that no subscription would be received for less than a half litre per second.

The company was also bound to lend to each and every land owner who subscribed for water, a sum equal to one hundred francs per hectare ($7 69 per acre) subscribed for, to be used by him in the preparation of his land for irrigation. This sum to be advanced in two parts, the first half on demand, and the second half three months after the first irrigation on lands prepared with the first half, and where it shall have been shown that the advance has been judiciously expended. The sums thus advanced were to be repaid in installments which, with interest, amounted to 6.25 per cent per annum for the fifty years, on the amount borrowed.

THE CANAL OF SIAGNE.[*]

A similar concession was made in 1866, also to the English company, for the canal of the Siagne and Loup, in the department of the Maritime Alps, under very similar conditions and for a like period to that governing the case of the Saint Martery.

In this instance the town of Cannes was to be supplied with water for domestic purposes, as well as the surrounding country to be irrigated, and it was made the co-grantee, to own the works in perpetuity, after the first fifty years when owned by the company.

---

[*] See, law of August 25, 1866, *Les Annales des Ponts et Chaussées*, Vol. LXXXVIII, p. 385, *et seq.*

The general government granted a concession of 500,000 francs ($100,000), and the town was to loan its credit to raise money for use on the works by the company, as in the case of the St. Martery, taking control of the revenue of the company from water rents as a basis upon which to capitalize for a loan, and being in turn assured from loss or embarrassment by the obligations of the company.

CANAL OF SIAGNOLE.*

This concession was made to five individuals forming a society, for fifty years, and afterwards in perpetuity to the department of Var. The society received a subvention of 30,000 francs ($6,000), and was authorized to derive from the Siagnole three hundred litres (10.6 cub. feet) of water per second, provided they at all times left at least a volume of water in the bed of the river such that the discharge might be one hundred litres per second above the dam of the manufactories of Mons.

The department was authorized to contract a loan, the interest of which, with all expenses connected with it, should not exceed three fourths of the amount of the rents of the canal, in order to aid the society in the construction of the canal. The total amount of this loan was not to exceed 90,000 francs ($18,000). The department was to receive the rents from the irrigators, and, after paying the interest and other expenses and installments of the principal, to hand the balance over to the society each year, until the debt should be paid off, when all the rents were to be paid to the society.

The works of the canal were declared to be of public utility; and the landed tax to be only for the simple amount of land occupied by the main canal and the secondary ditches, but the buildings and warehouses of the society were subjected to the usual tax.

The principal canal and secondary canals had to be entirely finished, and put in operation in the space of two years, counting from the decree of concession. The tertiary ditches, however, had only to be undertaken when the subscriptions would amount to six per cent of the expenses of their construction; but once begun, they had to be finished in two years.

The society was authorized to collect rents at the rate of forty francs per litre ($215 15 per cub. foot) per second for periodical waters of irrigation. The privilege of subscribing for less than a litre was given to the irrigators, but with the proviso, that for every quarter of a litre, or less than that quantity, there should be paid a rent of fifteen francs ($3).

---
* Decree of June 14, 1870. See, *Les Annales des Ponts et Chaussées*, Vol. CIII, p. 1206.

The irrigating proprietors were obliged to give free right of way under pain of not having the right to irrigate. Every proprietor who subscribed for a volume of water of twenty litres per second could have that quantity in a continuous stream by payment of the corresponding rent—this water to be delivered to him separately by a single gate. This same rule held good for a number of individuals, clubbing together to receive their water jointly through a separate gate.

The rents were to be fifty francs per litre ($268 95 per cub. foot) per second, if they were not subscribed for until after the decree of the concession, and the proprietors could free themselves from all rent charges by paying the capital, fixed at eight hundred francs per litre ($4,303 12 per cub. foot), provided they declared their intention so to do in the year following the decree of concession. Every time, however, the subscriber freed himself by depositing a capital, the society was obliged to deposit in the landed bank of France a sum necessary to constitute, by the accumulation of interest compounded for fifty years, the rent to be paid during the forty-nine years following.

In case of reduction or remittance of the rents from insufficiency of water, the year in which such reduction or remittance took place was not to count as one of the fifty years granted by the concession.

The company had to deposit 3,000 francs ($600) as security before the decree of concession was made, and this sum was to be restored to them when the expenditure on the canal amounted to 20,000 francs ($4,000), as certified to by the engineer-in-chief and the prefect of the department of Var.

The works of the canal were declared to be of public utility, and forfeiture was to be incurred for failure to construct in the given time.

In this canal we have an instance of a direct subvention to a society to assist in the construction of the canal, additional assistance in the shape of a loan authorized to be raised by one of the departments, and the transfer of the property, after fifty years, for the sole benefit of a department; also remission of taxes on the enterprise, except for buildings and land actually occupied.

THE CANAL OF THE BOURNE.*

The rights, privileges, and benefits for the canal of the Bourne, department of Drôme, hereinbefore spoken of, were conceded in 1874, for a period of 99 years, to three individuals for the benefit of a society or company, to be formed to carry out the project. The

---

*See, law of May 21, 1874, *Les Annales des Ponts et Chaussées*, Vol. CXXVI, p. 451, *et seq.*; also, De Passy, appendices 5 and 6.

district comprised 22,000 hectares (54,340 acres) of which 10,500 hectares (25,935 acres) were reckoned as irrigable, and a volume of 7 cubic metres (247 cubic feet) of water per second was allowed for the purpose of irrigation and other uses contemplated.

According to the terms of the law and agreement and schedule annexed, the government allowed the company an advance or subsidy of 2,900,000 francs ($580,000), which amount was one third of the estimated cost of the works inclusive of main, secondary, and tertiary canals and structures.

This subsidy was not finally granted, however, until after the company or society had been formed and subscriptions for water been secured to the extent of 3,000 litres per second at the rate of 50 francs per litre.

And the subsidy or advance was to be paid in installments, on completion of work in cost and value to the amount of three times the sum paid in each instance, one tenth of the whole being held back till the final completion of the work.

The company was to build all the works, and transfer them to the state at the expiration of the term of the concession, in good order.

The secondary canals and their tertiary branches may each go into the hands of a syndicate association of irrigators, for operation, in the sub-district served by it, should the landholders choose thus to organize and undertake the management.

The rate of water rents, for irrigation, was fixed at 50 francs per litre of flow for those persons who subscribed before the opening of the works, and 60 francs per litre for those who subscribed afterwards.

Supposing the entire volume of water conceded to be sold at the minimum figure, the revenue of the company would be (7 cubic metres=7,000 litres@50 francs) 350,000 francs per annum, which in fifty years would yield a return of 17,500,000 francs on the outlay to the company, which was expected to amount to 5,800,000 francs—being two thirds of the total estimated cost of the works, the government advancing the other third.

CANAL OF THE RHONE.*

This concession was made to three individuals acting for and in the name of a society then forming. The volume of water to be derived from the Rhone was 2,500 litres (88.3 cub. feet) per second in low-water. The concession to last for ninety-nine years, and the society to receive from the government a subvention of 900,000 francs

---
*Decree of August 7, 1878. See, *Annales des Ponts et Chaussées*, Vol. CXLIII, p. 531.

($180,000), provided the company could show subscriptions for at least 1,500 hectares (3,750 acres) of land to be irrigated.

The first installment was not to be paid until the society had expended 800,000 francs ($160,000). The three first fourths of the subvention to be devoted to the principal canal, the balance to be paid to the society on the provisional reception of the canal by government, with the exception of one tenth, which was to be paid after its final reception.

All the expenses of construction, operation, and maintenance were to be paid by the society, and it was to receive all incomes from the canal during the term of the concession, at the expiration of which the canal was to be returned to the state.

The society was authorized to contract one or more loans, the interest and expenses in connection with which were not to exceed 15,000 francs ($3,000). The first loan not to exceed 800,000 francs ($160,000), and no loan or issue of bonds to be made except with the authorization of the minister of public works, and after the entire subscription of the capital shares, and the employment in the works of four fifths of this capital.

The society engaged to execute all the works of the principal canal, secondary, and tertiary branches, as well as the works necessary to deliver to and carry away from the property of every one desiring it, the water for irrigation and domestic uses; to finish the main canal and put it in operation within the space of five years from the date of concession, and to complete all necessary canals within one year of the time in which they were commenced.

The annual rent was fixed at 40 francs per hectare ($3 07 per acre) for those who subscribed before the decree of concession, and 60 francs per hectare ($4 61 per acre) for those subscribing afterwards. The first subscribers had the privilege of afterwards augmenting their original subscriptions by an equal amount at the rate of 40 francs per hectare, but anything in excess of this amount had to be paid for at 60 francs per hectare; the right to the use of the water being in all cases inherent to the land and not to the individual.

The society had to give a security of 60,000 francs ($12,000) in cash, to be returned to them when they had expended on the works 200,000 francs ($40,000).

The works were declared to be of public utility.

The society agreed to pay to the state an annual rent of one franc, and the state reserved the privilege of revising this once every ten years.

In this concession we have an instance of direct assistance in the shape of a subsidy; the concession to last ninety-nine years, at the end of which time the enterprise is to be handed over to the state. The company was also authorized to contract one or more loans for its use in the construction of the works.

CANAL OF VESUBIE.*

This concession was given to the "General water company of France," under the direction of the engineer-in-chief of bridges and roads (the construction and management of these irrigation works being thus measurably a public work of the state), for a period of ninety-four years, and afterwards in perpetuity to the town of Nice. The enterprise was declared to be of public utility, and the General water company received a subvention from the state of 2,400,000 francs ($480,000).

The canal was to be taken from the river Vesubie, and so constructed as to carry at least four cubic metres (141.2 cub. feet) of water per second at the head of the first secondary canal, this discharge being fixed for the execution of the work, but not as a determinate quantity of water which had to be derived by the company.

The company had to pay all expenses of construction as well as all indemnities for temporary occupation or deterioration of lands and all damages resulting from the works.

All the expenses in connection with acquiring lands for the location of headworks of the canal and its dependencies, for modification, destruction, or stoppage of manufactories, for disturbances of users of water, had to be supported one half by the company and one half by the town of Nice, the company paying the town, however, 100,000 francs towards these expenses.

The indemnities due for the establishment of the tertiary canals and the distribution ditches for water, or for obtaining the passage of these waters over intermediate lands by right of simple servitude, had to be paid by the proprietors interested, who had to give proper titles for the same to the company.

The community of Nice granted to the company the lands which were required for the establishment of the reservoirs and their dependencies necessary for supplying the town with water, and the gratuitous disposition of all the ways of communication belonging to the community for the establishment of canals, ditches, conduits, etc., so long as such use did not interfere with their usefulness as means of communication. It also agreed to pay to the company an annual rent of 80,000 francs ($16,000), representing the municipal

---

* Law of December 26, 1878. See, *Les Annales des Ponts et Chaussées*, Vol. CXLIV, p. 1397.

subscription for a weekly delivery of 60,000 cubic metres of water. It was, however, understood that when the gross income of the company should amount to 180,000 francs ($36,000), municipal subscription included, any excess of income over this figure was to go towards the reduction of the annuity of the town, so as to limit it to 60,000 francs.

The company received authority to collect rents from the users of water, not in the town, as follows: For fifty centilitres, per second, 46 francs ($494 86 per cub. foot) per season of irrigation, for one litre, per second, 80 francs ($430 31 per cub. foot).

There were no subscriptions received for periodical water for a less amount than half a litre per second.

The buildings and storehouses of the canal and its dependencies had to pay the usual taxes, but the enterprise was only taxed for the actual amount of land occupied by it, reckoned as land of the first quality.

We have in this canal an instance of the construction and management of a canal largely at the expense of the state for ninety-four years, and then the transfer of it for the sole benefit of a municipality in perpetuity.

We have here, also, a remission of all taxes on the enterprise, except for its buildings and the land actually occupied by it.

THE PIERRE-LATTE CANAL.*

A recent instance of an enterprise of considerable magnitude to which the state made a large advance, and also guaranteed interest in a large amount, is the case of the extension and enlargement of the Pierre-latte canal, in the department of Vaucluse, the law for which was passed in 1880, making the concession to certain individuals for the benefit of a society to be formed to carry out the work.

The concession was for eight cubic metres (282.4 cubic feet) of water per second from the Rhone, to irrigate about 20,000 hectares (49,400 acres) of land, and for a term of ninety-nine years.

The estimated cost of the work, including main, secondary, and tertiary canals and works, was 8,000,000 francs, of which the State was to advance 2,000,000 on the work, in installments in amounts not exceeding one third the actual expenditure at any time according to detailed engineering reports.

In addition to this, the state guaranteed to the concessionary society undertaking to construct and manage the works, for a period of fifty years, a revenue of 4.65 per cent per annum on the remaining

---
* See, law of August 2, 1880, *Les Annales des Ponts et Chaussées*, Vol. CLI, p. 21, *et seq.*

6,000,000 of funds estimated to complete the work, and which the society was to raise for its capital.

This advance and guarantee, however, were not made until a certain revenue had been assured by subscriptions for water by the land-holding irrigators, so that the extent of government liability for interest was limited to 167,000 francs per annum.

And in case the income from the canal grows to be more than enough to produce the rate guaranteed by the state over and above the cost of operation, etc., one half this net revenue is to go to the state.

After the fifty years of the guarantee of interest, the state is to receive during the balance of the period of concession an interest of four per cent in return on the sum of the net amounts advanced by it as interest.

At the end of the ninety-nine years of the concession, the canal and all its dependencies are to become the property of the state.

The provisions with respect to the management and maintenance of the secondary and branch canals by a syndicate of irrigators in each case, were the same for this canal as for the canal of the Bourne.

THE CANALS MANOSQUE AND HERAULT.

A late instance of the state loaning money for the construction of irrigation works is the case of the Manosque canal, in the department of the Lower Alps, sanctioned by law in 1881.*

The canal was to take two cubic metres of water per second from the Durance river for irrigation.

The proprietors of lands to be irrigated were to engage to take water to a certain amount at a fixed rate, and, for fifty years, and to form themselves into a syndical association to manage the canal.

The state, thereupon, to advance all the money for the enterprise, amounting to two million francs, and to receive in return, during the period of fifty years, seventy per cent of the gross proceeds from water rents, which it was estimated would repay the state with interest.

The following is another instance of both a subsidy and a guarantee of interest on capital invested, brought into form by a law of 1882:

Canal of the Herault, department of the Herault, to take 3,500 litres of water at low-water, and 5,000 at time of flood, from the Herault river, for the irrigation and submersion of lands; to be built by a syndicate of landholding irrigators, at an estimated cost of 6,300,000 francs.

The state was to give a subsidy equal to one third of the total cost

---
* Law of July 7, 1881.

of the works, and to guarantee interest for fifty years at the rate of 4.65 per cent on 4,200,000 francs, the balance of money to be raised. This guarantee could only take effect after subscriptions had been made for water for 2,000 hectares of lands, at rates about fifty francs per litre.*

OTHER LATE WORKS.

In other instances the state has constructed irrigation works, and either turned them over to the landholders to manage, or reserved them for management as public works of the state.

An instance of the first above mentioned class of action, for which the law was passed in 1881, is that of the Canal Ventavon, in the departments of the Upper and the Lower Alps, and taking water from the Durance river.

The state granted the associated irrigators a water-right of 2,500 litres of water per second, on a nominal payment of one franc per annum, and then undertook to construct the main canal necessary to deliver the water, at a cost not to exceed 1,733,000 francs, estimated to be two thirds the total cost of the whole works, and to turn it over to the associated irrigators, for use forever, when they had built the necessary secondary canals and smaller ditches for distributing the water.§

Another instance of this kind of action on the part of the government is that of the canal de Petite-Vence, Department of Isére, taking water from the government canal Roize, and built for the irrigation and submerging of lands, at an estimated cost of 81,000 francs, of which the government was to expend two thirds on the main works, and the associated irrigators one third on the secondary canals and other works, the association to manage the canal forever.

Of irrigation canals, constructed on government account as public works, the following are of late dates:

Canal of the *Malpas*, department of Herault, taking water from the navigation canal of Midi, for the submersion of two hundred and ninety-six hectares, to be built by the state, at an estimated cost of 86,000 francs.†

Canal of *Saint Marcel*, department of Aude, taking water from the canal of Midi, for the submersion of three hundred and eighty-five hectares, to be executed by the state, at an estimated cost of 130,000 francs.

---
* See, law of July 13, 1882. *Les Annales des Ponts et Chaussées*, Vol. CLVIII, p. 1298.
§ See, law of July 20, 1881. *Les Annales des Ponts et Chaussées*, Vol. CLVII, p. 5.
† See, law of March 3, 1881. *Les Annales des Ponts et Chaussées*, Vol. CLII, p. 1263.

Canal of *Argeliers*, department of Aude, taking water from the canal of Midi, for the submersion of one hundred and five hectares of land, to be executed by the state, at an estimated expense of 80,000 francs.*

Canal of the *Raounel*, department of Aude, taking water from the canal de la Robine, and intended for the submersion of five hundred and three hectares of land, executed by the state, at an estimated cost of 320,000 francs.§

---

* See, law of August 17, 1881. *Les Annales des Ponts et Chaussées*, Vol. CLVII, p. 573.
§ See, law of September 22, 1880. *Les Annales des Ponts et Chaussées*, Vol. CL, p. 118.

AUTHORITIES FOR CHAPTER VIII.

In the preparation of this chapter I have consulted the following named authorities:

Dumont.—[Work cited as an authority for Chapter II (French).] See, Book II, Chap. VI, pp. 280–330.

De Passy.—[Work cited as an authority for Chapter II (French).] See, Chapter I, pp. 79–130.

Barral.—[Works cited as authorities for Chapter VII (French).] See, Chapters I and II of each volume, the descriptions of the several canals in other chapters, and the chapters relating to syndicate associations, canal companies, land proprietorship, and population.

Les Annales des Ponts et Chaussées.—[Work cited as an authority for Chapter III (French).] See, Vol. XVII, p. 289 et seq.; Vol. XLVIII, p. 523 et seq.; Vol. LXXXVIII, p. 162 et seq.; Vol. LXXXVIII, p. 385 et seq.; Vol. CIII, p. 1206 et seq.; Vol. CXXVI, p. 451 et seq.; Vol. CXLIII, p. 531 et seq.; Vol. CXLIV, p. 1397 et seq.; Vol. CLI, p. 21 et seq.; Vol. CLVIII, p. 1298 et seq.; Vol. CLVII, p. 5 et seq.; Vol. CLII, p. 1263 et seq.; Vol. CL, p. 48 et seq.; Vol. CLVII, p. 573 et seq.; and elsewhere in the publication.

IRRIGATION LEGISLATION AND ADMINISTRATION.

ITALY.

CHAPTER IX.—ITALY[1];

RIGHT OF PROPERTY IN AND CONTROL OF WATER-COURSES AND WATER SOURCES.

INTRODUCTION.—Importance of the study of irrigation experience in Northern Italy.

SECTION I.—*Ownership and Control of Water-courses and Waters.*
    Basis of property rights in water-courses in Northern Italy.
    Ownership: Lombardy; Piedmont. All Italy.
    Control: Lombardy; Piedmont. All Italy.

SECTION II.—*Ownership and Control of Springs.*
    Right of Property in Springs.
    Acquired rights to the use of Spring Waters.
    Regulation of the opening and use of Springs.

SECTION III.—*The Riparian Right.*
    In Piedmont, under the Sardinian Code.
    In all Italy, under the Italian Code.
    General remarks.

INTRODUCTION.

IRRIGATION IN NORTHERN ITALY.

The valley of the Po, in Northern Italy, is very generally regarded, and popularly spoken of, as the classic land of irrigation; and, indeed, if there is a region worthy of the name, these plains of Piedmont, Lombardy, and Venetia are deserving of it.

This valley is about two hundred miles in length, and varies from thirty to sixty in width, being bounded on the north by the Alps, and on the south by the Apennine range of mountains. Throughout its length, and keeping nearest the foot of the southern range, runs the Po, from west to east, a large river; while entering it, and joining this main drainage way, from the bordering mountain regions, are thirty or more other streams, of varied sizes and character; of which at least half a dozen are great irrigation feeders; and twice as many more contribute notably to the water-supply used in agriculture.

The valley is like our own of the Sacramento in size, and form, and disposition of water-ways, but is much better supplied with streams, and receives, on the average, about seventy-five to one hundred per cent more rainfall.

Irrigation was probably commenced by the Romans in this region, but the greater works of the country date since the tenth century; most of them were built after the fourteenth and before the beginning of the present century; while several notable ones, and the one of chief importance, have been constructed during the present generation. These works are constructed in the most substantial manner, with stone reveted banks in many places, and with masonry headworks, bridges, outlets, sluiceways, overfalls, syphons, and other structures. The volumes of water handled far exceed any conducted and distributed in this country, and the practice of irrigation is very much more refined in its details, than is our practice except in some notable instances.

The customs of the people of this region have crystallized into laws and regulations covering the whole range of points and subjects met with in the development of irrigation works and practice, so that we have here a rich mine of data in which we may find principles, and trace the working and results of principles, applicable to, and to be heeded in the formulation of the irrigation code of the future for California.

This development came to a point of completeness worthy of special attention, in the States of Lombardy and Piedmont, particularly, before the recent unification of the government of all Italy. So that I shall first trace as fully as necessary the systems of the Lombards and Piedmontese, and then present the law of all Italy, as it now exists, on the important points of our inquiry.

SECTION I.

OWNERSHIP AND CONTROL OF WATER-COURSES AND WATERS.

BASIS OF PROPERTY RIGHTS IN WATER-COURSES IN NORTHERN ITALY.*

For five centuries after the fall of the Roman empire of the west, the people of the Italian peninsula were tormented by successive invasions of barbaric tribes from different quarters. It becoming apparent that the ruling families of sovereigns of the various kingdoms, could not protect their subjects from pillage, the people concluded to protect themselves, and hence grew the spirit of inde-

---

* See, Sismondi, Smith, and De Buffon.

pendence upon which was formed the Italian republics of the middle ages.

Thus, during the tenth century the residents of the principal cities with the surrounding country, each organized an independent state with a representative form of government, and elected administrative officers. Forming leagues or confederacies at later dates, these states became republics, several of which, particularly on the seacoast and in the south of Italy, retained their independent existence with some vicissitudes of fortune, as late as the present century; but those in the north of Italy—the upper part of the valley of the Po, the quarter where irrigation has developed to the greatest extent—soon gave way to pressure of invasion from without, and to the machinations of local magnates, and the feudal system here made rapid progress to full development.

RISE AND FALL OF THE FEUDAL SYSTEM.

In northern Italy the independence of the feudal lords was most complete, and the hereditary principle was recognized not only as relating to the possession of local governing power, but to the possession and ownership of land. The counts took every means to oppress the allodial land proprietors who held titles from former rulers or under preceding forms of law. From such persistent and covert persecution, even the authority of the kings was often powerless to afford protection. Many private individuals voluntarily surrendered their allodial titles and consented to hold their property as the vassals of the counts, in order to get protection from the local potentate. Thus it came to pass that the feudal system of land tenure was established—none but persons of noble birth could hold property in their own right; all others held it as vassals of the dukes, counts, marquises, margraves, etc., and the land was known as the *feif* of the ruler.

The waters of all streams, which under senatorial and imperial Rome had been the common property of all the people, and the rivers, which had been the property of the sovereign power or nation, and which during the barbarian rule became the property of the rulers themselves, and then of the kings who followed them, and later of the people of the republics, now became the property of the local feudal lords.

The Roman laws had been lost to the people, and all records of them were at one time thought to have been destroyed; but among the unwritten laws of the country—in the customary law of the people, with respect to the management and distribution of waters in irrigation—were to be traced the influence of those principles which we find to have existed in the Roman system.

Documents of the tenth and eleventh centuries, recording and formulating previous practice, bear witness that "the principles of the Roman law in matters connected with the use of waters had never been wholly lost sight of, but, embodied in the traditions of the people, had continued in unwritten form to influence the development of agriculture. * * * The irruption of the barbarians brought into Northern Italy Germanic rights and the feudal laws. All the rights appertaining to the public centered in the feudal lord of a commune, a province, or a kingdom, becoming his absolute property. * * * It was not for purposes of police that the feudal superiors exercised all the rights of masters over the water-courses, but that their right of *absolute property* necessarily absorbed everything previously held to belong to the community. There existed, in fact, merely the relations of masters and subjects."—[Smith, Vol. II, p. 124; quoting Giovanetti.

"At the peace of Constance, in 1183, the Italian towns of the Lombard League recovered all the rights previously vested in the feudal superiors, and from that time the rivers have been held to be public property. These rights were then vested in the cities themselves, which each exercised authority over a certain extent of adjoining territory."—[Idem, Vol. II, p. 134.

FROM THE EARLIEST TO THE PRESENT LAW.

The earliest recorded laws of northern Italy date from the tenth century, when Otho the Great, emperor of Germany, granted the cities of Lombardy the right to live according to their ancient laws and local customs, which included their customs and regulations regarding irrigation.

A code of the republic of Milan, dated in the early part of the thirteenth century, contains an extended series of provisions regulating the use of water in irrigation, the right of way to conduct it in canals, and the privilege of diverting it from streams.

The laws of the republic of Venice, dated in 1455, recognize the ownership of running waters as being in the government as representing the whole people, forbid the diversion of water from the streams without "the requisite authority from competent magistrates," and provide that the waters may be used "by every inhabitant of the territory of Verona" "for the irrigation of his property," after obtaining the requisite authority and "under the condition that he inflicts no injury on parties possessing older rights to the same waters."—[Smith, Vol. II, p. 121.

When the monarchic element was introduced, there were constant struggles between the royal governments and municipalities on the question of the right to the running waters.

The result of these struggles was a recognition on the part of the governments of certain water-rights already utilized, but the successful assertion of ownership by them of all other waters. So that, to quote again the author above referred to: "In Northern Italy the waters of all streams, whether navigable or non-navigable, appertain to the royal or public domain."

OWNERSHIP: LOMBARDY, PIEDMONT—ALL ITALY.*

During a large part of the present century, and until 1865, the valley of the Po was under several separate governments, so that even the general laws were not uniform for all of this irrigation region, until a very recent date, and even yet regulations established by some of the local governments are still in force in the states for which they were promulgated.

In what will hereafter be said, reference will be made to the laws of Piedmont and to those of Lombardy, as they existed a few years ago, and until the merging of the governments into that of the kingdom of Italy, and then, for each heading, the provisions of the general civil code of Italy, known as the Code of Victor Emmanuel, and promulgated in 1865, will be given.

Lombardy.—That which was said in the final paragraph of the preceding section, had reference more particularly to Lombardy or the Lombardo-Venetian Kingdom, and as it existed under Austrian dominance.

The old established claim of the cities, communes, and associations of proprietors, and of noble individuals, to the supplies of water which they had for long periods of time actually utilized, having been recognized, the government asserted and maintained its ownership to all natural streams whether navigable or not.

Diversions of water under the old claims were subjected to government regulation, and no new diversions could be made, or new work built in the stream beds without special government authorization.

But when the government had come into full control of the streams, so many claims to their waters had grown up, that the propertyship of the state was almost a barren one, and it found itself heir to a struggle for the control of rights unregulated with respect to the public, and unadjusted amongst themselves.

Piedmont.—In the kingdom of Piedmont, also, the right of property in all running water was reserved to the state. This reservation applied not merely to the larger class of rivers, but also to the streams and torrents, the waters of which could only be used under specific grants from the government.

---

*See, Smith, Vol. II, Part IV, pp. 116 to 146, 248 to 263. The Sardinian Codes, and the Italian Civil Code.

A royal ordinance concerning the use of waters, and dated in 1817, commences with the following articles:

"I. All the rivers and torrents in the state are royalties, and by consequence they appertain to the royal domain.

"II. No one can establish channels or canals for the introduction of water into his property, either for the use of mills or other structures, unless he possesses a legitimate title to the same, or has obtained a royal grant."—[Smith, Vol. II, p. 248.

At a later date (1828) a royal instruction to the intendants of provinces, concerning the regulation of water-courses, commenced as follows:

"All the rivers and torrents in the state are *regali*, and belong in consequence to the royal domain.

"Hence, therefore, the sovereign permission is necessary before the waters can be used in any way whatever, either in agriculture or industry."—[Smith, Vol. II, p. 249.

The civil code of Charles Albert, of the kingdom of Sardinia, published in 1837, was to a very great extent a following of the Code Napoleon of the French, but in the matter of ownership of running waters, and water-courses, the preceding laws of the Lombardian kingdom are confirmed by Art. 420, as follows:

"The * * * rivers and torrents * * * and generally all those portions of the territory of the state which cannot become private property, are considered as dependencies of the royal domain." * * * The alienation or grant of such property as is specified in this article is subject to special rules, etc.

These rules were practically the same as others which preceded them, and made necessary the acquirement of special permits or concessions from government before water might be diverted from the streams for any purpose, except under the old established rights.

Of the principle here involved M. Giovanetti says: "We, in Northern Italy, have been judicious in ranking among the things appertaining to the royal or public domain, the waters of all rivers and streams, whether navigable or non-navigable. In this respect Art. 420 of our (the Sardinian) civil code is the reverse of Art. 538 of the Code Napoleon, which regards navigable rivers only as those belonging exclusively to the state."

The Kingdom of Italy.—After all Italy had been brought under one government, in 1865, was promulgated the civil code of Victor Emmanuel, of which Art. 427 is as follows:

"The national roads, the shore of the sea, the harbors, bays, coasts, rivers and torrents, the gates, the walls, the ditches, the bastions of forts and fortifications, form part of the public domain."

This provision of the code of 1865 is the law of Italy to this day, and under it all running waters, except those of very small streams, indeed, are claimed as the property of the government representing the people as a nation, and they are administered very much as are the waters of the navigable streams of the public domain of France.

Navigability, or only floatability for timber even, would not be a safe test for streams of great economical and public importance in Northern Italy, for the river beds are of such excessive slope and roughness, even where the volume of water is considerable and used by means of great works for irrigation, that navigation would be out of the question, except at very great expense for works of improvement. Although the rivers have been improved for navigation to some extent, works of this class have not been nearly so extensively prosecuted as in France; so that the streams at the heads of irrigation canals, although larger in point of volume of water than are the irrigation rivers of France generally, are frequently not navigated or even used regularly for the floating of timber.

We see, from these physical circumstances, an apparent underlying reason for the different definition of the public streams in Italy from that adopted in France. Navigability itself was a ruling consideration in France, while volume of water for irrigation was the point of importance which made the stream one of public utility in Northern Italy.

The codes of Charles Albert and of Victor Emmanuel say that "rivers and torrents" are dependencies on the public domain; and as a matter of fact, in Northern Italy every stream of perennial volume, other than very small streamlets, is regarded as a river (*fieume*); and every stream of intermittent flow from the rainfall or melting of snows, except the smallest, is regarded as a torrent (*torrenté*). Thus, it is only streams and ravines quite insignificant in size that are ranked as other than part of the public domain, and these are because the government has not chosen to extend the application of the words "river" and "torrent" to them to meet the requirements of the law in their cases.

GOVERNMENT CONTROL OF WATER-COURSES.

Under this heading will be given without further comment or remark, for the present, a number of extracts from various acts, laws, etc., showing the extent and nature of the control and management of water-courses which the recent governments of Piedmont and of Lombardy, and the present one of all Italy, have established or continued in force.

REGULATIONS IN PIEDMONT.

The first abstract is that of the *General Regulations for Water-Courses in Piedmont*, which were promulgated in 1817, and remained in force for some years at least, after the unification of the government of Italy, indeed if changed at all it is but quite recently.

And the second abstract containing some articles of the Sardinian penal code, applicable in Piedmont, and providing for the punishment of those who offend against government regulations and the laws respecting water-courses, and irrigation and drainage works.

(1) *General Regulations for Water-Courses in Piedmont.*

"All proprietors, possessors, or employers of canals, supplied by rivers and torrents, are forbidden to execute any works in the beds of the latter without the sanction of the authorities, under penalty of a fine not less than 10, and not greater than 100 *lire* (from about $2 to $20), in addition to the expense of replacing things in their original state, and of compensation for any damages which may have been caused to other parties.

"Proprietors, possessors, or employers of canals obtaining their supplies by means of fixed dams, are bound to maintain the positions and forms of these unaltered, to avoid raising their sills, or extending them farther across the beds of the rivers.

"When the supplies are obtained by means of temporary dams, made so as to be easily removed in times of flood, it is forbidden to render such works permanent, or to reconstruct them with heights, or in positions different to those previously in use.

"It is forbidden to proprietors, etc., of canals supplied either by permanent or temporary dams, to make any excavations in the beds of the rivers, whereby the supply would be unfairly augmented.

"Parties violating the foregoing provisions shall be bound to restore things to their former state, and shall, in addition, be subject to a fine not less than 100, or greater than 300 lire (from $20 to $60) for each offense.

"When changes in the condition of the streams may render alterations of dams or additional channels of supply necessary, the sanction of the superior authorities must be applied for. In such cases the claimant must lay before the intendant of his province a regular plan of the proposed works, prepared by a hydraulic engineer, and showing the part of the river and adjacent lands which will be affected by them, as also the different levels of the same.

"The intendant must visit the spot, or ascertain, through the agency of the government engineer of the province, that no injury to any one will result from the executions of the proposed works. All parties in the same, or in other provinces or districts, whose interests may be affected by the works, are to be heard for or against them, as may be.

"When, from unforeseen causes, want of water may arise, the proprietors, etc., of canals are authorized, in the event of urgency, to take measures to obviate the same, reporting their proceedings to the intendant of the province, who will cause the works to be inspected; and if they are found to be irregularly constructed, or likely to cause injury to others, will have them removed or altered as may be expedient.

"When changes in the course of the streams render works necessary, the matter shall be referred to the agency-general of finance: and the intendant-general, having obtained the opinion of the permanent commission (of engineers), will order the necessary proceedings.

"The proprietors, etc., of canals are bound to maintain all the works in an efficient state, and are personally responsible for any damages to others arising from their neglect.

"They are also bound to provide for the free escape of surplus water in time of flood, under penalty of a fine varying from 10 to 100 lire, in addition to giving compensation for damages.

"Siphons for the passage of waters belonging to private parties beneath the beds of streams, shall be maintained unaltered by their proprietors; and they are forbidden to execute any works connected with them, which might contract the sections or raise the beds of the rivers, under pain of a fine not less than 30 or greater than 150 lire, in addition to the expense of restoring things to their original state.

"Other articles prescribe conditions to proprietors of siphons under streams, binding them to permit these to be altered, as the government engineer may consider necessary, with reference to the protection of the public rivers."—[Smith, Vol. II, pp. 304 to 307.

(2) *Provisions of the Sardinian Penal Code—Applicable to Water-Courses, etc., in Piedmont.*

"ARTICLE 711. Whoever shall have voluntarily destroyed, cut, or broken through the dikes or embankments constructed for defense against the rivers, streams, or torrents, and shall have caused thereby an inundation in which one person has perished, shall be punished by death. If, however, this person has perished under circumstances which the offender could not possibly foresee, the punishment shall be that of hard labor for life.

"In every other case, the punishment shall be forced labor for certain periods, or, in lieu thereof, solitary confinement for seven years at least.

"ART. 712. If the destruction or rupture of the dikes and embankments, or like works alluded to above, shall be attributable to a simple fault, the punishment shall be that of imprisonment.

"ART. 713. As regards other breaches or injuries done or caused to dikes, embankments, bridges, hydraulic buildings, or other works of art, including such as belong to private parties, the punishment shall be that of solitary confinement. The tribunals may, however, in consideration of the circumstances and the nature of the injuries, substitute for the preceding, simple imprisonment.

* * * * * * * * * * * *

"ART. 718. Every individual who, without right or by means other than those above indicated, shall voluntarily cause waste, damage, or deterioration on the lands of others, whether by leveling or filling up ditches or canals, shall be subject to the penalties specified below:

"If the damage done shall exceed the value of 100 lire (about $20), the punishment shall be three months' imprisonment at least.

"If it does not exceed this value, the punishment shall also be imprisonment, of which the period may be extended to six months.

"In the two cases referred to above, the tribunals may add to the imprisonment a fine, which shall in no case be less than one half, or greater than twice the amount of damage done.

* * * * * * * * * * * *

"ART. 723. He who, without title, and without right, shall take water, or cause it to be taken from any reservoir, or from rivers, streams, torrents, rivulets, springs, canals, or water-courses, and shall appropriate it to any use whatever;

"He who, to the same end, shall break, or cause to be broken, the dikes, dams, sluices, or other like works, existing along the rivers, streams, torrents, rivulets, springs, canals, or water-courses;

"He who shall hinder, in any way, the exercise of rights which other parties may have acquired to the said waters;

"Finally, he who shall usurp any right whatever on the sources of water referred to above, or shall trouble any one in the enjoyment of the legitimate possession he may have acquired;

"Shall be punished by imprisonment, the period of which may extent one year; and by fine, the amount of which may be carried to 500 lire (nearly $100). The tribunals have the power of inflicting separately one or other of these punishments.

"ART. 724. If individuals possessing a right to obtain or use water, fraudulently cause their outlets, dams, or channels, to be constructed in forms different to those agreed upon, or having capacities of supply greater than those to which they have right, they shall be punished as guilty of abstraction of water.

"ART. 725. The proprietors, farmers, or other employers, who, in using their legitimately acquired rights to water, shall cause it to overflow the roads or lands belonging to others, shall be punished by a fine, which shall not exceed one fourth of the amount of the damage done.

"ART. 726. If the crimes contemplated in the present chapter shall be committed by the guardians of woods and waters, or by any other public agents, whose duty it is to check or prevent them, the punishment of imprisonment, when inflicted, shall exceed by one month, at least, and at most, by one third of its duration, the heaviest penalty inflicted on individuals not public agents, who may have been guilty of the same crime, provided always that the maximum of punishment fixed for the said crime shall not be exceeded."

REGULATIONS IN LOMBARDY.

The following abstracts are of regulations provided over a century ago for rivers and districts in Lombardy, and which were in force until quite recently, if, indeed, they are not so at the present time, with the addition, only, of a more complete administrative establishment for their enforcement.

These are regulations specially applicable to the river Lambro, the one dated in 1756 and the other in 1782, and both of them being republished under government direction in 1832:

(1) *Special Regulation for the River Lambro.* (1756.)

"The numerous disorders which exist along the entire course of the river Lambro, from its origin in the Lakes of Alserio and Pusiano, to its junction with the Po, having attracted the attention of the magistracy of the state of Milan, in consequence of the inconveniences and injuries at once to the royal treasury, and to public and private interests, which they have caused, most especially in the deficiency of water so frequently occurring, and traceable to them, and particularly as affecting the supply of the canal Martesana.

"The said magistracy, with the view of remedying such inconveniences, has judged it expedient, leaving in full force all former proclamations, especially such as affect the royalties of the waters, to publish the present edict.

"Whereby, in the first place, it is forbidden to every person, of every grade or condition, without exception, to divert the water of the river Lambro from its proper course. No one shall employ it for the irrigation of arable land or meadows, without the appropriate permission, and license by privilege or royal grant, under a penalty of three hundred crowns, of which two thirds shall belong to the royal treasury, and the remaining third to the guards of the river appointed for its protection, whose testimony, with that of one credible witness, shall be sufficient to warrant proceedings against offenders.

"All parties enjoying the use of water from the Lambro are warned against taking more than is secured to them by their respective rights, privileges, and grants, on pain of being proceeded against, not only for damages to the extent of the value of the water improperly taken in time past, but to entire deprivation of the water, and other penalties described in this edict; their outlets shall be closed, and the evidence of the guards, or any other parties reporting the offense, supported by a single witness, shall be deemed sufficient for conviction.

"It is forbidden to millers, or other parties possessing mills on the river Lambro, to retain or check the water in any way or under any pretext whatsoever. When the mills are not at work, the escapes shall be left open during the entire period of stoppage. Such mills as do not possess proper escapes, shall be provided with the same within eight days after the publication of the present edict, so that the water may flow freely into the bed of the river. These provisions shall be observed, under a penalty of one hundred crowns, to be applied as above described.

"Whoever, possessing the right to establish outlets or channels for the extraction of water from the Lambro, may have allowed the same to have become broken or out of repair, shall be bound to place them in good condition within one month after the publication of this edict, under the appropriate license of the magistracy, who will determine, according to the circumstances of each case, whether an inspection by the engineer or other official of the tribunal be necessary, or simply the assistance of the guard. If the repairs are not executed within the time specified, they shall be immediately afterwards effected under the orders of the magistracy, and at the expense of the recusants.

"It is forbidden to establish dams, or to construct works of any kind whatever, either across the bed or along the banks of the river, without the especial permission of the magistracy, under a penalty of two hundred crowns for each offense. The water shall be allowed to flow freely for the benefit of irrigators at lower levels, and particularly for the increase of the supply in the canal Martesana.

"All parties using the water of the Lambro are enjoined to obey the orders of the guards appointed to watch over the execution of the present edict, under a penalty of one hundred crowns, which may be increased at the will of the magistracy.

"Two guardians are appointed for the river, one having charge from the source, near the lakes Alserio and Pusiano, throughout the entire district of Crescenzago, and the other from this latter point to the junction of the Lambro with the Po. They are enjoined to watch carefully over the execution of the present and all preëxisting regulations, to secure for the river all the water that of right appertains to it, and to report all infractions of the orders of the magistracy, on pain of removal, and such other punishment as may appear due.

"No one shall be permitted to persevere in present or past abuses, on the plea of neglect, tolerance, or carelessness of the public agents. No such plea shall be accepted from any one in mitigation of punishment for breach of these orders; and the magistracy reserves to itself the power of taking whatever steps may seem to it best in each case, saving always such rights as may be vested in the royal treasury.

"This notification shall be published, not only in this city of Milan, but in the towns of Monza and Melegnano, and in the adjoining districts."—[Smith, Vol. II, pp. 187 to 190.

(2) Special Regulations for the River Lambro. (1782.)

"Retaining in full force all preëxisting regulations, and especially that under date the twenty-sixth July, 1756, the guard of the Lambro residing at Monza is enjoined to visit annually before the twenty-fifth of March, the springs, commonly called *teste* (the heads), by which the river is fed with the view of ascertaining that all these are well cleared, and that they really supply the entire quantity of water which could be obtained from them. All parties interested in such supply should depute persons to accompany the *camparo* during the said visits, to concert and arrange with him regarding the nature and extent of the necessary clearances, or of such other works as may be required for the efficiency of the heads. The guard should report the whole of these proceedings for the information of the magistracy.

"Having satisfied themselves of the correctness of this report, the magistracy shall order the execution of the repairs, the expense of which shall be recovered from the employers of the waters in proportion to their respective interests in the same. In addition to these expenses for works, a fair remuneration shall be fixed, at the discretion of the magistracy, for the assistance given by the guard.

"It is forbidden for the future to throw earth or rubbish, or other matter into the river, or to extract sand, except from collections of deposit; and in removing these, care shall be taken not to derange the natural level of the river. Excavations or ditches for the collection of sand or gravel are absolutely prohibited.

"If a necessity should arise for clearing earthy materials from the bed of the stream, parties desirous of doing so should communicate with the guard, who will satify himself that the work contemplated can cause no damage, either to the river itself, or to the adjoining properties. In the event of new work being undertaken, reference should be made to the magistracy, who will prescribe such conditions as may appear most appropriate in each case.

"Various sinuosities of the Lambro being caused by trees falling into the bed, or by spurs which throw the force of the stream on the opposite bank, to the injury of proprietors of land there, from the corrosion which is the consequence, the guard ought to immediately intimate to the owners of such tree or spurs that they must remove them within three days, otherwise they shall be removed by the guard himself, and all expenses for work or damage shall be at the charge of the proprietors.

"The trunks and roots of trees which come down the river in time of flood shall be removed by the guard; and as it is impossible•to know whose property they are, they shall be granted to him as a reward for his exertions in removing them.

"The soaking of flax in the river being injurious to the fish, it is absolutely prohibited; but parties may carry on the process, each in their own channels, and the guard should at once report any infraction of this order to the magistracy.

"To prevent any affectation of ignorance, his royal highness orders this edict to be posted in all public places along the river, and enjoins all parties to obey the agents of the magistracy."—[Smith, Vol. II, pp. 190–192.

SECTION II.

OWNERSHIP AND CONTROL OF SPRINGS.

CHARACTER AND IMPORTANCE OF SPRINGS.

The northern plain of the valley of the Po, throughout the very localities where the principal canals have brought their waters, is the site of a great number of *fontanili*, or springs, which afford a large and highly prized supply of water for irrigation.

Under extended areas of this plain, at depths from five to ten feet from the surface, lie beds of permeable gravel and sand filled with water, which the considerable transverse fall of the country puts under a slight head of pressure at localities towards the medium and lower parts of the sloping surface.

Doubtless many of these *fontanili*, like the *cienegas* of Los Angeles and San Bernardino counties in our state, formerly were natural springs or little marshes producing water, and have been developed and concentrated in their flow by artificial openings; but very many more, and their numbers mount up into the thousands, are purely artificial developments. They are made by digging into the permeable strata, and the waters, rising several feet, are brought out to the surface and on to the meadows further down the plain, by conducting them in ditches or closed conduits on grade slopes less than those of the country.

Besides these peculiar springs, which play so important a part in the irrigation of the plains, the country generally is one well supplied with subterranean waters, so that ordinary springs are plentiful, as in almost any region, upon the higher lands and in the hilly and mountainous districts.

Under these circumstances we might expect to find the recorded customs and laws of the countries replete with provisions touching the ownership and use of spring waters, and such is the fact, for there are veritable treatises of considerable length and intricacy, relevant to this subject alone.

RIGHT OF PROPERTY IN AND ACQUIRED RIGHTS TO USE SPRINGS AND SPRING WATERS.*

Lombardy.—The principle that ownership of the land carries with it all beneath its surface and all it produces, has prevailed from the times of the earliest recorded laws in all these north of Italy states. Waters rising out of the soil have always been regarded as the absolute property of the owner of the soil, so long as he retained them within the bounds of his estate, and did not permit his title to suffer abridgment by allowing some other proprietor to acquire a prescriptive right to the use of the waters.

The springs of the Milanese alone, in upper Lombardy, number upwards of seven hundred, and are frequently very valuable. Baird Smith tells of one, not an exceptional case at all, whose rising pool covered a space two hundred by one hundred feet in area, and which, supplying twelve cubic feet of water per second, was estimated to be worth $20,000.

The springs always remain the property of the owner of the soil, although the right to use their waters may be wholly alienated and held by the owner of some other property, either by sale or prescription. Baird Smith cites the following case:

"The irrigating water on this property was derived from a beautiful spring, which may be quoted as an illustration of the strange way in which rights of property to water have established themselves in this country. The proprietor could not tell me how or when the right of use was established in his family. No written record of any kind existed to prove it; but from time immemorial the use of the spring, though situated in the middle of another estate, belonged to the possessors of the land he held, and efforts made before the tribunals to invalidate his claim had entirely failed. He had, however, a right only to the water; to a passage for it and for his work-people along its banks; to sufficient space on each side of the channel for depositing the sand or gravel clearance; while the soil, trees, and produce of the banks belonged entirely to the proprietors of the farm on which the spring was situated."

Piedmont.—The Sardinian code of 1837 had the following provisions

---

* De Buffon, Vol. II, pp. 193-198; Smith, Vol. I, sundry places, and Vol. II, pp. 167-169, 254-257, and elsewhere.

with respect to the ownership and control of springs; and the acquirement or loss of right to the use of spring waters:

"ART. 555. He who has a spring on his land can use the same at his will, saving the right which the proprietor of the lower land may have acquired by title or prescription.

"ART. 556. The prescription in this case can be acquired only by an uninterrupted enjoyment during the space of thirty years, calculating from the moment when the proprietor of the lower land made and finished on the upper land visible works, designed, and which have actually served, to facilitate the descent to, and the passage of the waters through, his own property.

"ART. 557. The proprietor of a spring cannot change its course when the water necessary to the inhabitants of a commune, village, or hamlet, is obtained from it, but if the inhabitants have neither acquired nor prescribed rights to the water, the proprietor may demand an indemnity, which is regulated by the tribunals, on the report of professional men."

Remembering that this code was promulgated in 1837, about thirty-three years after the publication of the Code Napoleon, and that it was a codification from laws and decrees, some of them made and put forth for the country by Napoleon during the period of his domination of it, we readily appreciate the similarity of these provisions to articles 641, 642, and 643 of the French code.*

They are, indeed, in the original languages, worded, as near as can be, exactly alike, with the important exception noticed in the second couplet—articles 556 and 642. Taking advantage of the experience gained from the contests in France, occasioned by the uncertainty as to the location of the works which a proprietor must construct to facilitate the flow on to his estate, in establishing a prescriptive right to the use of spring waters, the framers of the Sardinian code evidently followed the decisions of the French courts noticed in chapter V, and which at that time had been full enough for guidance, and embodied in their code itself the explanatory provision whose absence had occasioned so much trouble in France.

They distinctly said that the works necessary in the establishment of the prescriptive right must be "visible works," and "*on the upper land*"—that is, the land where the water rises, and where it is owned—and that they must be maintained for thirty years. It is said that this provision has prevented a repetition in Lombardy of the long contests which troubled the French courts on this point.

The Kingdom of Italy.—The code of Victor Emmanuel,§ promulgated in 1865, for all Italy, and now the law of the country, presents

---
* See, Appendix I.
§ See, Appendix II.

in articles 540, 541, and 542, provisions corresponding to those of articles 555, 556, and 557 of the Sardinian code above quoted.

Article 540 of the new is identical in wording, in the original, with article 555 of the old code. The principle as to ownership of a spring is the same for all Italy as it was for Piedmont and other parts of the Sardinian kingdom.

Article 541 of the new differs in general wording from article 556 of the old code, as indicated by the translations given; and also contains the important addition to the effect that the works shall not only be "visible" and "on the upper estate," but shall be *permanent*, in order to constitute conditions to establish a right of use of the waters of a spring. Otherwise, the articles are substantially the same.

Article 542 of the new is differently worded, but has substantially the same meaning as article 557 of the old code, with the exception that the character of evidence required in the adjudication of damages, is left to other provisions of the law, and not specified for this case in the new code.

REGULATION OF THE OPENING AND USE OF SPRINGS.

Not only, from the earliest recorded custom touching this subject, has the ownership of ground-waters in Italy vested exclusively and completely in the owner of the land, but, within certain prescribed regulations as to distances from other works, every owner of lands might dig for water as he chose, and do with water so found as he saw fit.

The origin of these springs, scattered by thousands over the plain, being in a common water-bearing stratum, which was cut through by the natural, as well as cut into by the artificial surface drainage and supply channels—the rivers, creeks, and large canals—it was found at an early period in the development of the country that the opening of new springs drained the waters from old ones, as well as from the water-courses, when excavated too near thereto, according to circumstances.

Amongst the earliest of the statutes of Milan was one prohibiting the opening of a new spring on any property within a certain distance from the bank of any river, or within a certain other distance of any other spring already formed, under pain of a heavy fine to be forfeited to the treasury of Milan, and with the obligation to refill the excavation and extinguish the new spring.

Later legislation discontinued the prescribing of any fixed distance to be maintained between springs, but provided for leaving that point to experts to decide for each case according to circumstances.

REGULATION OF THE OPENING OF SPRINGS.

Lombardy.—The important parts of the legislation of Lombardy, regulating the opening of new water-courses, in force from the early part of this century to the consolidation of the kingdom of Italy, about twenty years ago, were contained in the law of 1804 and a decree of 1806. The item in point, of the law referred to, was as follows:

"ART. 55. It is forbidden to excavate or open springs, or heads of springs, water-courses, and channels, as also to deepen or increase the dimensions of excavations, or springs actually existing, in the vicinity of rivers or canals, within the distances which, according to the judgment of practical men, could lead to injury to the rivers or canals, or to their banks."

This law was one placing the running waters—rivers, streams, and torrents—under the charge of the public administration, and providing regulations to be observed in carrying out the charge. It did not relate, in any way, to springs and waters not of the public domain, except as might be necessary to protect the public waters. Hence we find that its provision concerning the distances to be observed in opening new springs, and making excavations which might cause the opening of springs, related to the "vicinity of the rivers and canals," only.

The decree of 1806 supplemented the above provision of the law, by the following paragraph, Article 12 of Title 2:

"Saving the prohibition in Article 55 of the law of 1804 [above quoted], it is permitted to every one to excavate springs on his own land, and to conduct the waters, respect being always had to any rights which other parties may possess."

It has been held that this provision of the decree of 1806 applied the rule of the law of 1804 to all excavations on private lands, regulating their distances from other springs, canals, etc., of private parties, as these had previously been regulated with respect to the location of public canals and rivers.

These rules were the result of a summarizing of the outcome of experiences wherein it was found that circumstances of soil, subsoil, and practice produced such great differences in the minimum distances to be maintained between new and old excavations and channels—these varying from 8 to 200 yards—that it worked hardship to establish any fixed distance, and equity could only be arrived at by a general provision of law, leaving its application to expert judges of the facts and natural laws in each case.

De Buffon says, with reference to the laws of 1804 and 1806:

"By this ruling, as may be seen, the legislator was compelled to adhere to the principle of leaving it entirely to the option of experts to fix the distances of new excavations from older established works, in the different localities, so as to cause no injury, without prescribing a minimum determined distance, as has been done in the case of the Piedmontese law. The fact is, that it has been very difficult to fix this minimum for a territory like the Milanese, where, in most any locality, one is sure to find water by excavating, and never knows but that it is water which has filtrated from some of the numberless canals which exist in the neighborhood."—[De Buffon, Vol. II, p. 228.

Piedmont.—The legislation of Piedmont on this subject was crystallized in articles 599, 600, and 602 of the Sardinian code in 1837, and in this form continued in force until the consolidation of the Italian kingdom in 1865.

These articles take the form of, first, prescribing rules for excavations, such as ditches, canals, etc., not designed for the purpose of opening new springs, and, then, applying these rules, with additions, to the case of excavations made expressly for the purpose of getting a new flow of water.

The articles concerning excavations for ditches will be given in a subsequent chapter of this paper. I state their main features here, and then give the article specially relating to excavations for springs.

In excavating upon one's own land, for a canal, ditch, or other similar purpose, the upper edge of the excavation had to be placed at a distance at least equal to its depth from the nearest boundary of the property of another; the face of the excavation had to be sloped away at a rate not steeper than one on one, and if local custom or regulations prescribed a greater distance or longer slope, then such custom or regulations had to be followed. And, furthermore, should the boundary of the estate be formed by a ditch or road owned in common, the excavation had to be at the distance mentioned, from the nearest edge of such ditch or road. These provisions are found in substance in articles 599 and 600; article 601 relates to the case where the line of boundary is formed by a party wall, or wall owned in common, and then comes the special provision concerning springs, as follows:

"ART. 602. Parties desirous of opening springs, of establishing heads or channels of discharge for the same, of making canals or watercourses, of clearing, deepening, or widening the beds, of increasing or diminishing the slopes, or varying the forms, shall be bound to observe such increased distances over and above that fixed in the preceding articles, and to execute such other works as may be considered necessary for the protection of preëxisting springs, canals, or water-courses, designed either for the irrigation of land or the supply of buildings.

"And in case of dispute between proprietors, the courts in deciding ought to aim at reconciling the respective interests of the parties in the manner most just and equitable, having due regard to the rights of property, to the advantage of agriculture, and to the special uses to which the water may be destined. And, further, in all cases where such proceedings may be necessary, they ought to determine and decree, in favor of one or the other party as may be right, that amount of compensation which may appear on grounds of justice and equity to be fairly due."

The Kingdom of Italy.—In the Italian code of 1865 the provisions above referred to and quoted from the Sardinian, were closely followed in tenor, so that it is only necessary to refer to articles 575, 576, and 578* to note the concurrence.

THE QUESTION OF DISTANCE, ONE FOR EXPERTING.

The ancient legislation of Milan prohibited the opening of new canals or spring heads within 66 feet of rivers, and 580 feet of pre-existing springs; that of Verona fixed the last distance at 639 feet; of Brescia at 106 feet; while the old laws of Mantua prescribed 24 feet as the minimum between new and old water-carrying or producing excavations of any kind; thus, illustrating the fact that, in different quarters, soils of very different degrees of permeability were found, and showing the necessity for leaving questions depending upon varying physical phenomena, to be determined as they arise rather than by any general rule of law; and explaining why the modern legislation of the country has provided for a proper experting of such cases and a decision of them on the facts and deductions properly due thereto.

The necessity for legislation of this kind is well presented by the following extract from the work of an Italian author, worthy of all attention on these subjects. He says:

" Agriculturists find it hard that they can scarcely strike a spade into their lands without running the risk of being summoned before the courts, and forced to give security against possible damages. The proprietors of springs and canals are wearied to death by having to remain always on the watch against the works undertaken by their neighbors, or of having to submit even to real injury from the difficulty of obtaining clear evidence of it."—[Smith, Vol. II, p. 249, quoting Giovanetti.

SECTION III.

THE RIPARIAN RIGHT.

Bearing in mind the fact of the definition of public streams as being "rivers" and "torrents," and that these words apply in fact to all water-courses except very small streamlets and minor ravines, we

---

* See, Appendix II.

may now go on to the consideration of the riparian water privilege accorded by the codes of Charles Albert (1837) and of Victor Emmanuel (1865).

Piedmont.—The Sardinian code (1837) contains the following provision:

"ART. 558. Any one whose land borders on a stream flowing naturally, and without the aid of works executed by man, and which has not been included among the rivers, streams, and torrents, declared in Art. 420 to be the property of the royal domain, may make use of it during its passage, for the irrigation of his property.

"Any one whose property is intersected by the same stream may make use of it within the limits of his own land, with the obligation, however, of restoring the water to its natural channel on its passing beyond the boundary of his estate.

"ART. 559. In the event of any dispute arising between the proprietors to whom such waters could be useful, the tribunals, in deciding, must conciliate the interests of agriculture, with, at the same time, a due regard to the right of property. And in all cases the local and special rules which regulate the course and use of the waters must be observed."

From this we see that the owner of one bank of a natural stream not considered of public importance, might make use of its waters in irrigating his riparian lands; and that the owner of both banks might also utilize it upon his estate, but that he had to return the waters to the natural channel.

This was a close following of the Code Napoleon, after which the Sardinian code was framed, and left open the question as to whether or not the owner of one bank had to return the water to its natural channel after use in irrigation, and if so, how much or what proportion he had to return. This, as we have seen, was a great question in France, which was, after long litigation, set at rest by decisions of the highest courts and rulings of the administration, declaring that the obligation to return the water to its natural channel applied only in the case of diversion for other uses than irrigation.*

The Kingdom of Italy.—In framing the Italian code§ in 1865, this ambiguity was done away with, somewhat, by the following wording:

"ART. 543. Whoever has an estate bordering on a stream which flows naturally and without artificial help, excepting such as are declared public property by Art. 427, or over which others have a right, may make use of it for the irrigation of his lands, or for the exercise of his industries, on condition, however, that he restores the drainage and residue of it to the ordinary channel. Whoever

---

* Refer to pp. 97 and 98, *ante*, and elsewhere. Remember that this applies to streams *not* of the public domain: In France, to streams not floatable for logs, even; and in Italy, those not of general consequence as irrigation feeders.

§ See, Appendix II.

has an estate crossed by such a stream may also use it in the interval of its transit, but with the obligation of restoring the drainage and residue of it to its natural course when it leaves his lands.

"ART. 544. Should a dispute arise between owners to whom the water may be of use, the judicial authority must reconcile the interests of agriculture and industry with the rights of property; and in all cases the particular and local rules applicable to the stream, or the use of the water, must be observed."

As the law now stands in all Italy, therefore, the owner of one or both banks of such a little stream may use its waters in irrigating his riparian lands, but he must restore "the *drainage and residue* of it" to the ordinary channel; while he who is not a riparian proprietor cannot take such waters at all without the consent of all of the riparian proprietors, nor can any one riparian proprietor assign his right to water from such a stream to any one else.

The riparian right to divert waters from a stream is confined to the case of very small streams, and is scarcely known in the Valley of the Po—certainly not on any of the streams which rank as important sources for irrigation supply.

On this subject Mr. Baird Smith has written as follows:

"Even the riparian proprietor is prohibited from using the stream which flows past or intersects his land, without the special permission of the government, both in Northern India and Northern Italy." * * * But "there are instances in both regions where, perhaps in remote places, in mountain valleys, or like localities, the running streams have been used for ages by the inhabitants without let or hindrance, or acknowledgment of superiority of any kind. * * * The framers of the Albertine Code,* wisely respecting rights founded in immemorial usage, include all such rights in articles 558 and 559, which seem to be most judiciously adapted to the peculiar circumstances under which these exceptions to a general rule have arisen."—[Smith, Vol. II, p. 256.

AUTHORITIES FOR CHAPTER IX.

In the preparation of this chapter I have consulted and compared the following named authorities:

Sismondi.—" History of the Italian Republics." By J. C. R. De Sismondi: 1 vol.; London, 1832.

Hallam.—" History of Europe during the Middle Ages." By Henry Hallam; 1 vol.; New York edition, 1853. See, Chapter III, " Italy."

De Buffon.—"Agricultural Hydraulics: Of the Canals of Irrigation of Northern Italy." By Nadault De Buffon, an engineer-in-chief of the Government Corps of Civil Engineers, France; 2 vols.; Paris, 1862. See, Vol. II, Chapters XXXVIII to XLVI.
[*Note.*—Although by the same author, this is a different work from that cited for Chapter II, and others succeeding, concerning French legislation, etc.]

Smith.—" Italian Irrigation: A report on the agricultural canals of Piedmont and Lombardy." By R. Baird Smith, captain of engineers, Benal Presidency; 2 vols.; London, 1855. See, Vol. II, Part IV, Historical Summary; Chapter I, Sec. 1, and Chapter II, Sec. 1; and elsewhere, as cited.

Sardinian Civil Code.—[See, authorities for Chapter X.]

Italian Civil Code.—[See, authorities for Chapter X.]

---

* The Sardinian code was promulgated by Charles Albert; and hence called "Albertine."

CHAPTER X—ITALY[2];

WATER PRIVILEGES AND CANAL WORKS, AND THE ADMINISTRATION OF WATERS AND WORKS.

SECTION I.—*The Right to Construct Works in and to Divert Waters from Streams.*
Governmental Policy in regard to Water Privileges.
Applications and Formalities for Water Privileges.
Terms of Water-right Concessions.

SECTION II.—*Administrative Regulation of Water-Courses.*
The Administration.
River Regulations.

SECTION III.—*Administration of Government Canals.*
The Administrative Bureau.
Canal Regulations.

SECTION I.

THE RIGHT TO CONSTRUCT WORKS IN AND DIVERT WATERS FROM STREAMS.

GOVERNMENTAL POLICY IN REGARD TO WATER PRIVILEGES.

During the times of the ownership of the streams and waters by the sovereigns of the states, and by the petty feudal rulers, and by the sovereign powers of the states as the representatives of all the people, in each case, as has been spoken of under preceding headlines, the right to divert water from any river or torrent could only be acquired in the states of northern Italy by special grant or concession of privilege made on a formal application, after due examination and consideration of all the interests to be affected, and all the circumstances likely to affect the interest acquired under such grant.

And now that the country is united under one government and the waters belong to the royal or public domain, the same rule and substantially the same formalities in applying it exist.

Milan.—The earliest recorded laws of any of the northern Italian states—the Milanese code of 1216—contained an express prohibition of the act of building a dam or other structure in the channel or bed of a stream without due authority, and prescribed a process necessary to be gone through with in obtaining such authority. This principle of active governmental control and administration of the streams is found in all the compilations of laws which follow, for the region of the former republic of Milan.

Venice.—During the fifteenth century the republic of Venice promulgated anew throughout its irrigation provinces, regulations as to diversion of water from streams, similar in principle to the laws of Milan. Those thus published for the province of Verona commence with this declaration:

"Every inhabitant of the territory of Verona is at liberty to derive, from the rivers appertaining to the state, such supply of water as is necessary for the irrigation of his property, on obtaining the requisite authority from competent magistrates, and under the condition that he inflicts no injury on parties possessing older rights to the same waters."—De Buffon, Vol. 2, p. 297.

Having said thus much for two of the ancient governments, we come now to those of modern times in these regions.

Lombardy.—It appears that the policy of the rulers in Lombardy until the later years of its existence as a separate state, has generally been to dispose of the waters of its streams in absolute property, by gift or sale, to those who constructed the canals to lead them out, or itself to lead them out in canals and sell them directly or indirectly through "farmers of the canal revenues," to the irrigators.

One notable exception to this rule was during the first domination of the Austrian government over the Lombardo-Venetian provinces; at which time a regulation for the administration of matters pertaining to water-courses was issued, which contained this clause:

"In making grants we do not thereby vest in the grantee the right of property in the water, but only the right to use it either in irrigation or for hydraulic works. The right of property shall remain as heretofore among the rights appertaining to the crown."—[Smith, Vol. II, p. 212.

At the period of the consolidation of the Lombardian kingdom, such a great number of rights to water had grown up and called for recognition, that the waters left at the disposal of the state were reduced to a comparatively small quantity.

"In exercising its right of property in these waters for irrigation, the government of Lombardy followed one of three courses: 1st. It disposed of the water in absolute property, to parties paying certain established sums for it. 2d. It granted perpetual leases of the water on the payment of certain sums annually. 3d. It granted temporary leases for variable times at certain annual rates, the water reverting to the State on the termination of the lease."—[Smith, Vol. II, p. 135.

The first named course in policy was most common in the earlier years of the existence of the government; and, at that time, the last named plan was the least often resorted to.

At a later period the policy of granting the water in absolute property was almost abandoned; that of granting perpetual leases became prevalent; and the third method of granting temporary leases came into favor. And these two courses were those followed by the Lombardian government at the time of the consolidation of the Italian government, and the extinction, as an independent power, of that of Lombardy.

Piedmont.—The government of Piedmont has generally been more conservative in the care of its waters than that of Lombardy. Absolute grants of ownership of waters ceased in that country before the beginning of the present century. Water privileges for all time have been indeed issued, but the full right of regulation was reserved to the government, and the session of propertyship in the water was expressly disclaimed.

This reform, however, occurred of late years as compared to the origin of many water rights in the country, and the important works have absolute rights of ownership in the waters acquired in the centuries gone by.

During the later years of the existence of the Piedmontese government its waters were disposed of only on long term leases, drawn up with great care and in minute detail.

The Kingdom of Italy.—This last mentioned policy is that pursued by the government of Italy since it has supplanted those of Lombardy and Piedmont; the duration and terms of concessions being, as we shall see, quite similar to those already written of for France in the chapters of this report which have gone before the present.[*]

APPLICATIONS AND FORMALITIES FOR WATER PRIVILEGES.

Piedmont.—The acquirement of water privileges in Piedmont and the operations of diversion were, for many years previous to the consolidation of the Italian government, regulated by the following royal "Instruction to the governors of provinces and the agents of the royal domain, with respect to grants of water from rivers and torrents," dated in 1828:

"* * * Sundry statutes and patents formerly published, have hitherto regulated the provisions for grants; but as it is desirable to establish one uniform rule of procedure in such cases, the secretary

---

[*] Letters from Hon. George P. Marsh.

of finance, whose duty it is to obtain the royal sanction to proposed grants of water, has decided that in future the following orders shall be observed:

"I. Parties desirous of obtaining grants of water from the royal rivers and torrents, whether for irrigation or the movement of machinery of any kind, must present to the intendant of the province where the head of the proposed derivation is situated, petitions addressed to his majesty and authenticated by the signatures of the petitioners, or by those of a notary and advocate.

"II. To each petition the undermentioned documents should be attached:

"(1st). A regular plan of the locality, on which shall be noted the works which it is proposed to construct in the bed of the river or torrent, and the adjacent ground, so far as it may be connected with these works.

"(2d). Longitudinal and transverse sections of the river whence the supply of water is obtained, marking thereupon the depths in time of flood, and under ordinary circumstances; also, the height of the works to be established in the stream, and of the head of the ditch.

"(3d). A detailed report, proving the utility of the proposed works, and that they cannot cause any injury, either to other parties, or to the river, or torrent itself.

"These documents must be prepared by a hydraulic engineer. But in the event of no hydraulic engineer being near at hand, or of the works being of limited importance, it is permitted, but with special reserve, to employ a civil architect, or land surveyor, in the preparation of the papers above referred to.

"The intendants of provinces will render all practicable assistance to parties interested, so as to enable them to comply with the rules of the superior authorities.

"III. The petition and the documents above specified should all be prepared on stamped paper.

"IV. The intendant, on receiving the claim and its appendices, shall satisfy himself of their regularity, and shall depute the official engineer of the province to visit the spot at a specified time, to investigate the practicability of the project, and the propriety, or otherwise, of carrying it into effect; as also to decide on whatever precautions or modifications regard to public or private interests may require.

"V. The visit must be preceded by a publication of the claim, within the limits of the district specially interested in it.

"If the claim and the works proposed are in any way connected with the interests of more than one district, the notification should be made at the same time throughout the whole.

"VI. The order of the intendant should contain a brief summary of the nature and extent of the proposed works, and an invitation to all parties interested in them to be present at the time appointed for the visit, when they can explain their views, either verbally or in writing.

"VII. The report of the official engineer ought, in all cases, to furnish full and clear details on the following points:

"(1st). On the quantity of water to be taken from the river and the special use to which it is to be applied.

"(2d). On the form and dimensions of the headworks to be constructed, being careful to note that the provisions expressed in article 16 of the regulation of the twenty-ninth of May, 1817, are vigorously to be enforced.

"(3d). On the directions, heights, lengths, forms, and mode of construction of the dams required to raise the water.

"(4th). On the precautions to be observed by the grantee, when the supply is to be obtained by means of temporary dams, in replacing the same after the floods. Grantees being generally inexpert and careless in hydraulic operations, a matter so important as this proceeding should not be left dependent on their wills, but definitive measures should be prescribed whereby the injuries likely to be caused to the beds of rivers or torrents by badly constructed dams may be guarded against.

"(5th). On the capacity and slope of the canal for the passage of the water.

"(6th). On the means to be adopted to insure the regular execution of the works, to restore (when such is possible) the water to the stream at a lower point, and to protect all parties from damage by overflow of the canal or otherwise.

"(7th). And, finally, the official engineer ought to detail any local peculiarities which may have influenced his opinion.

"With such information before it the permanent commission of engineers (to which the project will be referred) can better decide on the propriety of sanctioning the final execution.

"The various documents above referred to will be attached to the royal patent authorizing the grant, in order that both the administrative and the judicial authorities may always have the means of ascertaining precisely the terms of the said grant, and of restricting the grantee within the limits of the same.

"VIII. On the receipt of all the papers connected with the case, the intendant should forward the same to the agency general of finance, with his own opinion upon them.

"IX. So soon as the agency general receives notice from the secretary of finance that the royal patent for the grant has been signed, it will communicate without delay with the intendant, who will transmit the information to the official engineer, to the syndic of the district, and to the petitioner, requiring the latter to procure the aforesaid patent from the secretariat of finance, and to pass it through the offices of the agency and the chamber of accounts, within the space of four months, under pain of forfeiture.

"X. The receiver general shall be supplied with the necessary instructions to enter the patent on his list, and to arrange for the collection of the annual water rent."—[Smith, Vol. II, pp. 249-253. See, also, De Buffon, Vol. II, p. 223, *et seq.*

Lombardy.—Several regulations of a like tenor prescribed the forms of application and proceedings to be observed in obtaining water privileges in Lombardy, but their provisions are so like those of Piedmont, just transcribed, that it would be a useless repetition to give them here.

The Kingdom of Italy.—When these north of Italy governments were set aside in that of unified Italy, much of the machinery of the hydraulic administration in the valley of the Po was retained. Regulations were continued in force temporarily, at least, and thus the old established forms and local rules were, many of them, still in application as late as 1882, and, it is believed, stand as laws to this day; although at that time a movement had recently been made, and was still on foot, to set them aside for a uniform and general regulation on each subject for all Italy. But the principles in the proposed new water code are substantially a transcription of those in the old rules so far as these, for the different localities, could be reconciled to each other; and, hence, we may look upon the old regulations cited as being substantially those of to-day.*

TERMS OF WATER-RIGHT CONCESSIONS. LOMBARDY—PIEDMONT.

Lombardy.—Previous to the recent consolidation of the Italian government, the general terms of water-right concessions in Lombardy were fixed in a regulation dated in 1806, and which was in this particular as follows:

Water-right regulations—1806.

Title I. Diversions of Water from Rivers, Torrents, and Public Canals.

"Art. 1. No one can divert public waters nor employ them for mills without a concession from the government.

"Art. 2. This grant specifies the quantity, the duration, the manner, and the conditions of the derivation, and the particular use of the waters, and establishes the annual rent which corresponds and is due.

"Art. 3. The terms of the preceding articles are not intended to work prejudice against actual possessors in their rights and uses for the water-heads and mill-rights to which they already have just title under the terms of the laws and customs in force in the different provinces.

"Art. 4. No new grant can be made to carry injury to existing rights. These will be protected, by appropriate reservations, from the influence of later concessions. To this end all petitions (for new grants) are published and posted, engineers are consulted, and together with their reports the proper conditions for the conduct of the work are inserted in the regulation.

"Art. 5. It is prohibited to change, under any alleged right, the actual state of outlets and of fixed dams without the permission of the government.

"Art. 6. The works made for diverting water by the aid of movable dams must be approved by the engineer-in-chief of the province, who must give notice thereof to the direction-general.

"Art. 7. The engineers are charged to take care, in the public interest, not only of the use of the waters conceded for irrigation and for mills, but that the clauses and conditions imposed in the ordinances are observed.

"Art. 8. To this end, they must keep in their offices a register, in which are recorded all concessions.

---

* Letters from Hon. Geo. P. Marsh.

"Art. 9. In case any one having a right to use water commits any abuse (of the right) the engineers-in-chief are authorized, by virtue of their office, to reëstablish the place in its original state and under their direction; for this power must be fully expressed in all the acts of concession.

"Art. 10. When contests occur concerning the use of waters, devoted to no other object than the interest of individuals, they shall be tried, as of old, by the ordinary tribunals.

"Art. 11. When, in such contests, public and private interests are both concerned, they are to be carried before the administrative authority."—[De Buffon, Vol. II, pp. 226, 227.

Piedmont.—By the terms of the Sardinian code, applying to Piedmont, grants for the use of water from streams of the royal domain were made only on condition that no injury should be brought about to legitimate rights previously acquired.

In the construction of works and management of waters under such grants, the grantees were obliged to avoid backing up waters upon those holding rights above them, or precipitating waters in undue volume on those below them, or depriving others of the waters which was due them. And should any damage accrue from their acts of omission or commission, they were bound by the terms of their grant to repair the same, and further to suffer such punishment as might be provided by the local or general police regulations.

In conducting their waters under such grant they were obliged to construct works according to prescribed and approved plans, to maintain those works under government supervision, and to observe the regulations provided for the ruling of such matters.

The following are the articles referred to :

"Art. 631. The grants for the use of water appertaining to the royal domain are always made on condition that they involve no prejudice to anterior and legitimately acquired rights to the same water.

"Art. 632. Parties having the right to extract and divert water from rivers, streams, torrents, canals, lakes, or reservoirs, are bound to avoid injuring those situated above or below them respectively, by the stagnation or by the backing up, or by the change of course of the said water. Whoever by neglect may cause any damage in these ways, shall be bound to repair the same, and further to suffer such punishment as may be established by the regulations of the rural police."

The Kingdom of Italy.—Upon this point, articles 614 and 615 of the Italian code* contain the provisions corresponding to those above from the Sardinian code.

The terms of water right concessions under the Italian government

---

* See, Appendix II.

will be more fully brought forward in a subsequent chapter under the heading "Irrigation enterprise—Cavour canal."

SECTION II.

ADMINISTRATIVE REGULATION OF WATER-COURSES.

THE ADMINISTRATION.*

The administration of water-courses and waters in Piedmont is already sufficiently explained in the provisions of the "Instructions to intendants of provinces," transcribed under the subhead of "Applications and formalities," given in the first section of this chapter.

The organization and system of the Lombardian government was so nearly identical with that of Piedmont in this respect to render unnecessary any detailed reference to it here.

It now remains to glance at the present system for all Italy, which indeed was founded upon that of Piedmont.

The Kingdom of Italy.—The executive functions of the Italian government are exercised by ministers appointed by the king. Amongst these are a minister of public works, and a minister of agriculture, industry, and commerce.

As in the French administrative organization, there is a bureau of civil engineering attached to the ministry of public works, but the organization is not so broad or complete, nor the employment of the engineers so general throughout the country, in the guarding of the streams and waters and the regulation of works, as in France.

But, for the valley of the Po, the systems of the Lombardian and Piedmontese governments have been perpetuated, so that there is in this great irrigation region almost as complete an organization as that already described for France.

There have of late years been several movements to reorganize the public works and engineering service for all Italy, but from various causes these have not been consummated. There is, however, a general and permanent hydrographical commission, composed of civil engineers of the hydraulic service, which supervises all affairs connected with water-courses and water-rights, and the minister of public works is himself a civil engineer of high attainments.

There is a special hydraulic service too, as in the French system, and all applications for water privileges have to be considered as much

---

* Letters from Hon. Geo. P. Marsh; also, see, Encyclop. Brit., Vol. XIII, pp. 448-464.

at length and in detail, and more particularly from the engineering, technical, and physical points of view, and less from those of the law and local sentiment, than in the case of the French system.

Thus, the engineers are made the judges of the local necessities and public advisability, or utility, and report directly to the central administration, and upon a broader view of each proposition than the French engineers are required to. While the local administrative officers are called upon for their opinions separately.

LOCAL ADMINISTRATIVE ORGANIZATION.

This local administration is made up as follows: Under the government as now organized, the valley of the Po is embraced within the departments of Piedmont, Lombardy, Venetia, and Emilia; and they comprise twenty-eight provinces. These provinces are the real administrative units, each being presided over by a prefect as is the case of the departments of France. The provinces are divided into communes, and each commune is presided over by a chief magistrate called a syndic.

The prefects and the syndics are appointed by the king, and there are provincial councils and communal councils associated with these officers respectively, as in the case of the French departmental administrative system. But unlike the French organization, the communal unit has direct communication with the central government, and is really the important factor in the ordering of internal affairs.

In general terms, therefore, we find the prefects of the provinces and the syndics of the communes charged with the administration of the affairs of the water-courses locally, in so far as the policing of the stream and the enforcing of regulations are concerned, but the engineers and the ministry of public works regulate the construction and maintenance of works in the streams and the diversion of water from them.

This, of course, relates more especially to the streams of the public domain. But it is to be remembered that in Italy all streams of any importance as irrigation feeders are public, and that, except on insignificant water-courses, and those remote from the centers of irrigation, or in mountain valleys, there are no claims of right to the waters or to the channel beds, founded on riparian proprietorship.

There are, however, some streams controlled altogether by associations of landholders or canal and water right owners, and over which the government has only a supervisory duty based on the ground of police power. But these rights are founded on ancient special grants of proprietorship in the waters and channels, and not on the ownership of the bank lands.

With respect to administration, then, the communal and provincial officers are the chief local executive functionaries in care of the policing of streams, generally, to carry out the regulations which emanate from the central government; and the engineers are a distinct branch of the administration, having to do with the question in their separate class of duty.

ADMINISTRATIVE WORKING—RIVER REGULATIONS.

The regulations under which the affairs of the water-courses of the valley of the Po are administered, are largely of origin in the first half of the present century, and after the formation of the Piedmontese and Lombardian governments of that time.

The principles involved are quite similar in them all, and it is only necessary to give one example here, in addition to what is incidentally said relative to this subject under other sub-headings, to sufficiently present the essential features of the system and the spirit in which it finds its motive.

ADMINISTRATION OF WATER-COURSES—PIEDMONT.[*]

In Piedmont the water-courses and royal canals were in charge of an administrative organization known as the *Agency of the domain*, the subordinate employés of which were river-guards, apparently corresponding in general duty to those of France, heretofore written of.

The instructions to the various "Agents of the domain" filled a large octavo volume, and went into great detail. Articles 357, 358, and 368 provide in effect that the class of agents of the domain to whom they are particularly addressed should guard the rivers and streams, watching for infringements of the regulations concerning diversions of waters and building of structures in the channels, aiding those who observed the laws, and reporting those who transgressed, to the director of the domain, and, after obtaining a warrant, proceeding to their arrest and the enforcement of the law concerning the establishment of things in their original state.

The agents, say these instructions, ought to be continuously on duty, for water is a thing which men are prone to take without due authority and to the grave injury of their neighbors and the public, and stream channels easily receive serious injury from thoughtless building in them.

These agents also are enjoined to be thoroughly acquainted with

---

[*] See, De Buffon, Vol. II, p. 214.

the laws and regulations touching water-courses and their duty connected therewith, and to know well the character and extent of rights which people have on the streams within their districts.

The domain receiver in each district is charged with the duty of seeing that works are constructed under concessions or grants in conformity to the terms thereof, and that they are properly maintained according to the opinion of the engineer. And he must report to the director of the domain all that is worthy of attention from that officer.

REGULATIONS FOR WATER-COURSES—PIEDMONT.*

The affairs of rivers and torrents of all classes in Piedmont were subject to regulation under a decree, of 1817, of which I present an abstract, as follows:

Navigable Rivers.—All persons were prohibited from diverting waters from navigable streams, and from placing any structure in a channel of any such stream, under a penalty of $2 to $30, and also the obligation to remove it and restore things to their former condition.

Old dams, for whatever purpose used, could not be changed or repaired without administrative permission and supervision, under pain of a similar penalty and obligation to restore them, etc.

Trees and underbrush growing along the banks could not be cut, except by administrative authority and inspection; nor could any clearance be made for cultivation within a distance of about 350 feet from each bank, without due authorization after inspection, under pain of a penalty of $2 to $20.

Owners of alluvial lands along rivers or torrents must keep their cultivations at the prescribed distances therefrom, or coming within those distances must have a permit for such action, from the intendant of the province, guided by the advice of the communal council, and of the provincial engineer. Penalty for infringement, $6 to $40; together with destruction of the plantation.

The digging of wells or opening of streams within certain distances of the banks of streams was prohibited. Penalty, $20 to $60.

Owners of bank property were, under regulations and by permits, allowed to protect the banks from washing. But revetments of masonry, brushwork, or other protecting constructions, must in no case project into the channels, except as these might be planned and executed under the supervision of the provincial engineer.

The intendant of the province, on the advice of the engineer, had

---

* See, De Buffon, Vol. II, pp. 314–319.

immediate direction of these matters, and there was an appeal from him to the direction-general, which acted on the advice of the central commission of engineers.

Non-Navigable Streams.—All persons were prohibited from diverting waters from or placing any structure in the channel of any non-navigable stream, ranked as a stream of public utility or importance.

For permission to divert water from such stream, or erect any work in its channel for the purpose of using its water or protecting its banks, application must be made to the intendant of the province.

The intendant directed an examination to be made by the engineer, as a preliminary to all permits, and interested parties were notified to meet the engineer on the ground, and make any desired representation to him about the project.

The only difference between the treatment of cases on these streams and those on navigable ones was in the form of proceedings and permits.

The free flow and open channel of small streams must be preserved.

Bank owners might, on due authority, construct works to protect the banks, but, if calculated to arrest the currents, or deflect them injuriously against either bank, they were removed.

The management of the details of the affairs of such streams was intrusted to syndical associations of proprietors interested.

A provision inserted in all grants of right to water, or right to construct works in a stream, was that the proprietors should constantly keep the weirs of the dams open, to leave ample space for the passage of flood waters.

For offenses against these regulations concerning non-navigable streams, similar penalties were imposed to those specified for like infringements of the rules applicable to navigable rivers.

Old dams and structures for diversion of the water, or for applying it in use, in any way, must not be changed in form, dimensions, or elevation, without due permission issued after examination.

The channels of these streams must be kept clear to a standard width, fixed for each stream in each commune, at the expense of and by the riparian proprietors, under direction of the provincial engineers.

The banks of these water-courses might be cultivated, but neither roots nor branches might encroach on the bed of the stream.

Islets could not be cultivated or cleared, except at the permission of the intendant of the province.

When such water-courses had low banks, subject to overflow, the riparian proprietors were under obligations to keep the channels clear of deposits down to the normal elevation for the bed.

Consumers of water, or users of it for power purposes, were called on for a share of these expenses of such maintenance of channels.

GENERAL RIVER REGULATIONS—LOMBARDY.

The following general regulations for water-courses for the province of Mantua, made while wholly under Austrian dominion, and continued while a part of the Lombardian kingdom, will convey a good idea of the general policy and extent of power in this respect exercised by that government:

River Regulations for the Province of Mantua.

"ART. 1. The damming up, directly or indirectly, of water-courses of any class or kind, or the alteration of any escapes, weirs, or channels, in such manner as that the water may be turned to the use of the offending party, or to the injury of others, is prohibited, under a penalty for each offense of 2,000 *lire* (upwards of $300), of which, half shall be granted to the informer. Failing payment of the fine, the offender shall be sentenced to imprisonment with hard labor for one year.

"ART. 2. The chief sources of injury to the banks, and of obstruction to the free course of the waters, are trees, underwood, or bushes of any kind. It is forbidden to plant these on the banks of the public canals and rivers, and such as exist shall be cleared away within twenty days from the publication of this edict. After this time the wood shall be cut down by the public officers, and sold for the general benefit of the associations of the rivers and canals.

"ART. 3. The lines of piles placed in the channels to facilitate fishing cause serious damage. These shall all be removed and sold for the general benefit; and, in future, whoever replaces such works shall be subject to a fine of 100 *lire* (about $15), whereof one half shall be granted to the informer.

"ART. 4. The proprietors of mills and their work-people are forbidden to raise the water, by any means whatever, above the levels either already fixed, or to be fixed hereafter. During floods, they shall be careful to open the escapes, so as to prevent damage. Each offense against this rule shall subject the offender to a fine of 200 *lire* (about $30).

"ART. 5. All proprietors of ditches shall be bound to maintain them in thorough repair, so that no water may escape from them into the public roads, or in any way cause damage to other parties, under a penalty for each offense of 200 *lire*, in addition to payment of all expense for injuries done.

"ART. 6. All employers of water shall obtain the quantities defined and fixed by their titles and grants. Forfeiture of all right to water shall follow the illegitimate alteration or extension of the prescribed areas of irrigation.

"ART. 7. Like forfeiture shall be the consequence of any improper interference with any of the various kinds of works on the canals. When a change in these is desired application shall be made to the magistracy of water for the province, who will order the proper steps to be taken.

"ART. 8. Employers of water who have irrigated the areas assigned to them, shall be bound to allow the surplus waters to flow off freely for the benefit of lower lying lands. To this end every proprietor shall be bound to establish drainage channels for the collection of the surplus waters; and neglect in doing so shall entail forfeiture of all right to water from the respective canals.

"ART. 9. [Orders that periodical inspections of the canals be made by the prefect or vice-prefect of the province, so as to insure observance of the provisions of the edict.]

"ART. 10. It being a common but mischievous practice for parties to carry water to lands so placed that the surplus waters are entirely lost, it is ordered that every landed proprietor shall cause to be made, at his own expense, a map of his property, on which the irrigable land shall be shown in its true dimensions, and with its heights above the sources of supply of water clearly exhibited; also, all the water-courses, culverts, roads, or principal canals, aqueducts, weirs, locks, and every other kind of works, shall be plainly shown. This map shall be preserved as a record in the office of the magistracy of waters, and shall be corrected from time to time, as changes are duly sanctioned by the proper authorities. Neglect of the present order shall be punished by loss of rights to the water.

"ART. 11. No changes of any kind shall be effected but under the orders of the magistracy, executed by the prefect or vice-prefect.

"ART. 12. The conservators of the different irrigating associations are enjoined to watch over the efficiency of the works under their charge. They shall make an annual inspection, and submit a report on the works to the congregations of their respective associations, indicating all the repairs or new structures required, and estimating the probable expense thereof. The visits shall be made during the first days of the month of February, and the congregation shall be held about the middle of the same month. By which means all needful repairs may be completed about the middle of April, when the demand for water arises.

"ART. 13. The conservators shall be careful to clear the canal beds of all water plants and weeds, causing them to be dug out by the roots for some distance from the water's edge, throwing the refuse clear of the embankments. If necessary, clearances of this class shall be executed three times a year.

"ART. 14. All parties are enjoined to receive, and execute with promptitude and good will, the orders of the conservators of the different associations. Disobedience shall be punished by a fine for each offense of 200 *lire* (about $30), which shall be increased at the discretion of the magistracy; if any offense be committed a second time by the same party, it shall be lawful to proceed against him under the provisions of the municipal laws.

"ART. 15. The annual tax shall be paid by all parties within the time prescribed by the congregations, and defaulters shall be proceeded against without further notice.

"ART. 16. Parties not possessing legal rights to irrigation shall not use, even to the smallest extent, the waters of the canals. The first offense against this rule shall be punished by a fine of 1,000 *lire* (about $150), with forfeiture of all the irrigated produce, and compensation to parties injured by the misappropriation of the waters. The second offense shall be punished by the confiscation of the land illegitimately irrigated.

"ART. 17. We reserve to ourselves the right to make grants of water for irrigation; and we hereby declare, that if it should come to our knowledge that arable or forest or meadow lands have been broken up for the purpose of creating rice-fields, in excess of those fixed by considerations of public police, and duly limited thereby, the grants thus abused shall be revoked; and we give notice that we will not in future allow any new rice cultivation to be established, until it has been proved to our entire satisfaction that the lands to be so employed are all in such low lying localities as to be unfitted for use under any less injurious kind of cultivation.

"ART. 18. In making grants, we do not thereby vest in the grantee the right of property in the water, but only the right to use it either in irrigation or for hydraulic works. The right of property shall remain as heretofore among the rights appertaining to the crown.

"ART. 19. In all grants for the use of water whencesoever derived, from colature or from works, we maintain in full force the provisions of existing agreements, in consideration of the benefits hitherto derived from their observance.

* * * * * * * * *

"ART. 26. To insure the reform of abuses, and to protect the interests of the royal treasury, all employers of water shall be bound to submit their titles, after due notice, to a deputation of officers, which from time to time shall visit the canals, with full authority to investigate and dispose of all cases brought before them, according to their judgment.

"ART. 27. The guards and police shall use all diligence in protecting the interests intrusted to them, and shall denounce all contraventions to the secretary of the magistracy of waters. In cases of neglect, the offending party shall be declared incapable of again serving the state; but if collusion or participation be established, he shall be sentenced to imprisonment with hard labor for a period not exceeding three years, according to the decision of the magisterial chamber.

"ART. 28. The magisterial chamber shall determine all farther provisions necessary to the execution of our laws, and shall decide on all matters connected with the waters of the province."—[Smith, Vol. II, pp. 208, et seq.

SECTION III.

ADMINISTRATION OF GOVERNMENT CANALS.

THE ADMINISTRATIVE BUREAU.[*]

In Piedmont, and also in Lombardy, the greatest irrigation works were the property of the government.

Some of these great canals date from very early times; indeed, their origin is quite obscure, except that it is known about when they were built.

These works were maintained under the supervision of government engineers, but, as a general thing, their revenues were farmed out in

---

[*] See, Smith, Vol. I, p. 120; also, De Buffon, Vol. II, pp. 218-220.

bulk to some contractor or association, who received the waters at certain outlets from the main distributaries, in large volumes, undertook to distribute them to the consumers, collect the revenues, and pay the government certain fixed sums annually for the privileges. This financial system was open to and resulted in great abuses, but with that phase of the question we are not concerned here.

There were also on these canals certain old water-rights, conceded by former governments to consumers, for some consideration or service rendered in years or centuries long gone by. Some of these rights were free from rate paying, while others were subject to an annual payment, generally at low rates.

But, however the waters were distributed, or under whatever right of use or rate of payment, the works were the special charge of government engineers, and their maintenance, extension, and remodeling contributed to develop a service of unprecedented skill in hydraulic construction and science.

Piedmont.—In the preceding section of this chapter I have spoken of the care and regulation of public streams in Piedmont through the services of the agents of the public domain. It now remains to speak of the management of the public canals. These are in reality great artificial public streams from which private canals draw, and considerable populations are supplied.

Their maintenance and general management was committed to the care of the ministry of finance, as a separate trust from that of public works generally, which were in charge of a minister of public works. The fact that the canals were a property yielding a revenue to the state in which the finance was more interested than any other bureau, is advanced as the reason for this arrangement.

Attached to this ministry of finance was an office of works, which was the executive agency in charge of construction and maintenance of the canals. The general management of the department was intrusted to the intendant-general of finance, the chief executive officer under the minister himself, but the *personnel* of the service was almost exclusively made up of civil engineers, of whom there were about twenty, together with their assistants and subaltern helpers.

The duties of this corps were connected entirely with the professional and practical labor of construction, maintenance, and operation of the works. The financial management and care of distribution of the waters were under the control of the contractor or farmer of the canal revenues, who ordinarily leased the waters in bulk for a period of nine years, and then sublet the water privileges.

Thus, there was an entire separation between the executive management of the canal works and the business management of the canal operation.

Under the engineers and their assistants there was a subordinate organization of guards, or superintendents and overseers, composed of one chief and thirty-five ordinary guards, whose duty it was to take local charge of the works. These were generally men of experience in the management of canal works, and they lived in houses, built for the purpose, close alongside of their sections of duty. To them were intrusted the keys of the distributing gates from the main canals, and hence they were persons of considerable importance, and not infrequently became skilled as practical hydraulicians.

Articles 359 to 367 of the "Instructions to agents of the domain," spoken of in the last section, contain provisions regulating the financial relations between this establishment and the lessees of the waters of the canals.

Articles 630 and 631 provide for the duties of the engineers in connection with the maintenance of main outlets for distribution, the expense of which was to be borne by the lessees of the waters in each instance.

Other articles prescribed in great detail the duties of the "agents" and of the "engineers" who were the officers of the two lines of administrative operatives under the intendant-general of finance. Of these duties, it is noteworthy that each agent and each engineer was required to keep a daily journal in detail of all his official actions and observations, according to a prescribed form, and to return such journal in duplicate with a summarized statement in the form of a report, also in duplicate, to the intendant, monthly, who retained one copy and transmitted the other to the intendant-general together with his observations. In addition to this, quarterly financial reports were required from all officers or agents in charge of works, and professional reports on the condition of works, from the engineers, also every three months.

The state, through the medium of this establishment, administered, maintained, and operated the canals, giving out the waters to the branch distributaries whence they were measured out, and the rents collected by the employés of the farmer of the revenues, as elsewhere spoken of. This system of farming the revenues to an individual, or individuals, was done away with in 1854, when all the waters thitherto thus disposed of were leased to the "Association of irrigation west of the Sesia," as is explained in a subsequent chapter; and the system of maintenance and operation of the works by government employés was also done away with by the leasing of the canals themselves to the

Cavour canal company, in 1862, also spoken of in detail hereinafter; and, finally, the management of the works, again by the general government of Italy, upon the failure of the Cavour canal company, remains to be mentioned.

GOVERNMENT CANAL REGULATIONS.

Piedmont.—Returning to the times of Piedmontese administration of the royal canals in the upper part of the valley of the Po, to carry forward the subject in a complete manner, I transcribe the following draught of "Regulations for the administration of the royal canals of irrigation," under which the works were managed until turned over to the Cavour canal company, and which constitute one of a number of regulations incorporated into the "Instructions to the agents of the domain," heretofore mentioned:

Regulation for the Administration of the Royal Canals of Irrigation.

Of the maintenance of canals.

"ARTICLE 1. All the royal canals of the kingdom are subjected to the present regulation.

"2. The general control of the royal canals is vested in the agency-general of finance, the executive duties being performed by engineers and guards appointed by it.

"The latter, with the guards appointed by the farmer of the canal revenues, shall take an oath of fidelity in presence of the judges of their respective districts.

"3. The engineers and guards are charged to prevent all interference with the waters, works, and employers of the canals.

"4. The articles of the regulation of the twenty-ninth of May, 1817, are maintained in full force.

"5. No one unprovided with a legal grant or right can make any use whatever of the canals; and any interference with the free course of the waters in the main channel or branches thereof is forbidden. Violations of any part of this article shall be punished by a fine of from 50 to 150 *lire* (from $10 to $30) for each offense, in addition to compensation for damages.

"6. Parties having a legal grant of water, but taking more than the quantity they are entitled to, or using at a different hour from that specified in the agreement among the employers of a common channel, or violating in any other way the terms of their grants or agreements, shall be subject to a fine of from 50 to 100 *lire* ($10 to $20), in addition to compensation for damages.

"7. Whoever shall raise or lower the gates of the outlets or escapes, alter, break, or deface the chambers of the works of measurement, force the locks of the same, or change their dimensions, shall incur a fine of from 150 to 300 *lire* ($30 to $60), in addition to the amount payable for damages. When the offense is perpetrated on crown property, the pecuniary fine shall be accompanied by imprisonment for a period varying from one to six months.

"8. Employers of the canals shall maintain their irrigation outlets and channels in forms prescribed by their grants, under a penalty of from 50 to 100 *lire*.

"9. The water flowing from irrigated lands, commonly called *coli* (*colatura*), shall be permitted to enter the canals freely, except when special agreements to the contrary have been entered into, under a penalty of from 50 to 100 *lire*, in addition to the price of the waters intercepted.

"10. It is forbidden to fish in the canals, or to excavate sand from them, or to use boats on them at any time, under a penalty of from 10 to 30 *lire*.

"11. The agency-general of finance may permit fishing, navigation, or excavation of sand, having first procured the opinions of the engineer and the director of the domain. Such permission ought to indicate clearly the portion of the canal to which it applies. It can be granted only for a period of not longer than one year, and is null and void unless registered by the grantee at the office of the direction of the domain, and of the local secretariat of the province.

"12. It is forbidden to establish, without the authority of the agency, bridges, fords, or ferries, and, also, to cross the canals, either on foot or with cattle, under a penalty of 10 *lire*, in addition to the expense of destroying works executed in contravention of this article.

"13. (Repeats the above with respect to minor works.)

"14. Whoever takes possession of land along the canals which belongs to the royal domain, removes the landmarks, makes excavations, carries away the produce of the plantations, or traverses the banks with cattle, carts, or conveyances of any kind, shall incur a fine of from 5 to 10 *lire* for each offense, in addition to the repair of any damages which may be caused, or to the cost of the things carried away.

"15. The possessors of land fronting or adjacent to the canals are forbidden to open new springs, to excavate ditches, to form ponds, water-courses, or channels of any kind, within a distance of 200 metres (nearly 220 yards) from the said canals, except in such cases as may be specially decided upon by the engineers, who will then fix such distances as may seem to them sufficient to prevent any leakage of the waters of the canals into the works referred to.

"It is also forbidden to the aforesaid possessors of land to plant trees within a distance of 3 metres (about 3⅓ yards) from the boundaries of the canals. Infringements of this article shall be punished by a fine of 10 *lire*.

"16. It is forbidden to cut the trees on the canal banks, or to carry away the prunings of the same, under penalty of a fine equal to double the value of the trees or prunings. If the trees cut and carried away shall exceed the value of 25 *lire* (about $5), the offender shall be imprisoned for not less than one month, in addition to paying the fine as above.

"17. Parties acquiring by legitimate titles any right to the plantations along the canals, shall not be allowed to cut or prune them except at the times and to the extent specified by the engineers in charge.

"18. All parties are forbidden to pasture cattle on the banks of the canals at any period of the year, under a penalty of from 1 to 3 *lire* for each animal."—[Smith, Vol. II, pp. 307-310.]

Lombardy.—In Lombardy the organization and regulations affecting the government canals was substantially the same as in Piedmont, so that there would be nothing added to the useful data of this report by introducing here anything specially relating to this branch of our subject for that country.

Italy.—Upon the unification of the Italian government, all the public canals of Lombardy and Piedmont, not leased to the Cavour canal company, were given over to the charge of the ministry of public works for all Italy.

Something will be seen of the management of a portion of these in a subsequent part of this report, and as it is substantially that followed by the Piedmontese government, with the exception of the different and broader organization of its administrative department, it is unnecessary to refer in detail to it here.

AUTHORITIES FOR CHAPTER X.

In the preparation of this chapter I have consulted and compared the following named authorities:

De Buffon.—[Work cited as an authority for Chapter IX.] See, Vol. II, B. VII, Ch. 39, Divs. I and II; Ch. 40, Div. I; Ch. 41, Div. I; B. VIII, Ch. 45, Div. I; and elsewhere as cited.

Smith.—[Work cited as an authority for Chapter IX.] See, Vol. II, P. VI, Ch. I, Secs. I and V; Ch. II, Secs. I and V; and elsewhere as cited.

Sardinian Code.—"The Civil Code of the Kingdom of Sardinia." Edited and annotated by A. Boron, Advocate, etc., 2d ed. Turin, 1857.

Italian Code.—"The Civil Code of the Kingdom of Italy." Edited, annotated, and compared with its predecessors, by Domenicantonio Galdi, Advocate, etc.; Roy. 8vo., pp. 1,400; Naples, 1865.

Ency. Brit.—Encyclopædia Britannica. Ninth Edition. Article, "Italy."

Letters, etc.—Letters from the late Hon. Geo. P. Marsh, U. S. Minister to Italy; dated at Rome and Florence in 1882, and addressed to the writer hereof, in answer to letters of inquiry on the subjects of this report.

CHAPTER XI—ITALY[3];

REGULATION OF IRRIGATION PRACTICE.

SECTION I.—*Distribution and Measurement of Waters.*
    Hydraulic Science and Practice.
    The Problems of Distribution and Measurement.
    The Piedmontese Legislation—Sardinian Code.
    Remarks on the Sardinian and Italian Codes.
    Distribution by Volume, by Use or Service, and by Time.

SECTION II.—*The Rights of Irrigators.*
    To a Continuance of Water Supply.
        The Right in Piedmont.
        The Right in Lombardy.
    To the Use of Spare Waters.
        The Sardinian Code.
        The Italian Code.

SECTION III.—*Obligations and Rights of Irrigators and Canal Men.*
    Obligations Concerning Water Supply and Use.
    Piedmont, Lombardy; all Italy.
    Priority of Privilege in Distribution.
    Piedmont; all Italy.

SECTION I.

MEASUREMENT AND DISTRIBUTION OF WATERS.

HYDRAULIC SCIENCE AND PRACTICE.

Until within very recent years, when there has been much activity and emulation in the perfecting of means and methods for the economical and exact measurement and distribution of waters in irrigation in France, in British India, and also on some special works in Spain, the works and regulations designed for the consummation of these ends in northern Italy have stood alone as evidences of an attempt at the systematic application of scientific principles to the details of an extended and complex practice of the art of conducting and measuring water in open channels for irrigation.

Commencing in the centuries that have passed, hydraulic science developed with the advance of irrigation and drainage practice in

Italy. For a long time this was its repository; and out from this country it subsequently spread.

"With the revival of knowledge in Italy, the art of hydraulic engineering was called into existence, and the extensive demand for skill in its details created early a supply of men familiar with all of these. Hence the remarkable number and great talent of the executive engineers, by whose exertions, rewarded and stimulated by their wealthy and powerful employers, that vast network of irrigation channels was spread over the entire surface of the country."—[Smith, Vol. II, p. 135.

The physical, social, and political conditions of northern Italy alike contributed to the growth of this science: The difficulty of tapping the chief sources of water supply, except by means of great works, requiring skill and technical knowledge to plan and construct; the necessity, produced by climatic and hydrographic circumstances, for making these works most substantial, and, consequently, costly; the complexity of the natural water supply system, and the confusion as to water claims which had grown up; the absence of system in the earlier works and projects; the consequent extreme complexity of works; the great value of water in irrigation; the wide destruction of property occasioned by waters of floods; the alarming unsanitary results of unskillful irrigation, insufficient drainage, and injudicious embanking of lands; and the natural outgrowth of confusion and litigation which resulted, made the necessity for men at once in command of scientific knowledge and practical skill in hydraulic work.

In following out the systemization of irrigation works and practice in that country, not only have the main works for the diversion and conducting of waters been in charge of those trained and educated to the task, but the practical studies and applications in the most minute details of distribution and measurement of waters have been equally committed to the care of specialists.

"Under this system, it is astonishing to see the extent to which minor canals have been executed. The whole surface of the country is covered by them as by a dense net-work. At all levels, and by the use of various ingenious works, they pass over, or under, or through each other, in such a way as to preserve individual rights uninterfered with, though the result to outward appearances, is a system of such marvelous complexity as to make the observer conclude it must lead to interminable disputes."—[Smith, Vol. I, p. 41.

We should, hence, expect to find, and we do find, that government itself has done much towards the advancement of knowledge and skill in this practice. In Piedmont, for instance, not only was there an establishment of civil engineers in the employ of the government and in charge of all public works, and having supervisory duties connected with water-courses and works relating thereto, but the study and private practice of the profession itself was the subject of state solicitude and aid.

"The economical importance of irrigation in Piedmont has naturally induced the government to furnish all practicable facilities for its study. The education of the hydraulic engineer is conducted with care, and no one is allowed to practice the profession without having graduated regularly at the university of Turin."—[Smith, Vol. I, p. 12.

That government established and continuously maintained stations for experimenting on and observing the flow, measurement, and distribution of waters, which were attached to the educational institutions and made accessories to instruction in hydraulic science, so that an education as an hydraulic engineer was, in that country, eminently practical as well as theoretical in its course and results.

The civil engineers were graded, according to their attainments, as hydraulic engineers, civil architects, and surveyors or land measurers, and no one not specially qualified for the higher rank of hydraulic engineer was permitted to practice that branch of the profession.

Such the men to whom were confided the works of irrigation in Lombardy; such the care with which men for this service were trained in Piedmont; and now the Italian government equally encourages the hydraulic art and science by means similar, and, hence, the details of the Italian system and the rules of practice and principles of law attending and governing that practice, are well worthy of study.

THE PROBLEMS OF DISTRIBUTION AND MEASUREMENT.

Next in order to those great complications and contentions which with irrigation enterprise are developed between governments and the grantees or employers of water for irrigation, between different grantees or employers, and between these and riparian proprietors, come questions which grow out of the relations between those who have water to distribute and those who want it to use—between the canal owners or managers and the irrigators.

Here are encountered the problems of equitable distribution and accurate measurement.

Water is contracted for and delivered in irrigation under three general systems of reckoning: the first, delivery to irrigate any certain crop or area of land for the season or for the time; the second, delivery of some certain quantity of water; the third, delivery of some certain flow of water for a certain period of time.

The contentions which arise and the sources of dissatisfaction with results, to both the canal or water man and the irrigator, under each of these systems, will form the subject of a chapter in another part of this report, so that it is sufficient simply to call attention to them here as being the moving cause of much solicitude and study in all well settled irrigation regions.

Irrigators generally, in the older irrigation countries, prefer the system whereunder they can have measured out to them a fixed quantity of water at certain periods of time, and then have the liberty to do with it as they choose. The difficulty of accurate measurement, under the very many and varying conditions attending the delivery of water, in new countries prompts and often makes necessary the adoption of the other systems.

In the measurement of waters two distinct ideas are to be held in view. These are: what unit of measure is to be taken; and, what means of measurement are to be adopted.

All civilized countries have a system of weights and measures applicable in the meting out of ordinary merchantable commodities and lands, but few have any established system for the measurement of waters. Such a system grew up in northern Italy, or, rather, several such systems found birth and development in the various provinces or petty states of the valley of the Po.

These systems were far from perfect, as we may view them now from the standpoints of an advanced hydraulic science, but they served a most useful purpose, and were the starting points from which irrigation engineers have sought to advance in other countries.

We find the laws and regulations of irrigation referring to certain standard measures and measuring apparatus, and, in view of what has preceded, we are prepared to appreciate their meaning, without going further at this time into the definite interpretation of these standards to those of our own country.

THE PIEDMONTESE LEGISLATION—SARDINIAN CODE.

In Piedmont water was distributed under three systems of delivery. The *first*, according to the quantity stipulated in actual volume; the *second*, according to the use, or the area to be irrigated; and the *third*, according to the time or season for which a flow was engaged. These different methods of delivery necessitated as many types of agreement, and each gave rise to its class of questions, so that legislation was demanded by which to regulate the contests that were brought about.

And hence we find in the Sardinian code the articles which here follow, and which were the outcome on the principal points of the experience theretofore had in Piedmont.

Articles of the Sardinian Code.

"ART. 641. In future when an agreement shall be entered into for a constant and determined quantity of running water, and the agreeing parties shall settle between themselves the form of the outlet or structure of derivation, then that specific form only shall be retained. The parties concerned shall not be permitted to impugn its correctness on the ground either of excess or deficiency of supply, unless such difference in either way shall exceed one eighth of the quantity agreed upon; and the action shall be instituted before the expiration of three years from the time when the work was first brought into use; always excepting the case in which the increase or deficiency of water may arise from changes in the supplying canal itself, or in the volume of the water flowing in it.

"If, in the absence of any agreement for a specific form, the outlet in actual use shall have been peaceably possessed and employed for the space of ten years, no complaints regarding either excess or deficiency of water shall be entertained, excepting in the case of variations in the supplying canal, or in the course of the water flowing therein, as above specified.

"In default of any agreement regarding the form of the outlet, or of possession, the form shall be determined by the tribunals, on the judgment of professional men nominated by consent between the parties, or if they cannot agree, by the tribunals themselves.

"ART. 642. When grants of water made for a specific service or object, do not express in terms the quantity granted, they shall be held to accord that volume which is necessary for the fulfillment of the said service or use. It shall be lawful for the parties interested therein to fix, at any time, the form of the outlet, and so to limit it as that the grantee shall receive the volume sufficient for the service agreed upon, but nothing more.

"When, however, the parties shall have agreed to give a definite form to the orifice of discharge or the outlet, or, in default of an agreement, there shall have been a peaceable possession of such form for the period above defined, objections to the same shall be admitted only in the cases and within the periods established in the former article.

"ART. 643. In new grants of water wherein a constant quantity of running water shall be agreed upon and specified, the said quantity shall be expressed in all public acts in terms of the 'module of water.'

"The module of water is that quantity which, under simple pressure, and with a free fall, passes through a quadrilateral rectangular opening, so placed as that two of its sides shall be vertical, with a breadth of two decimetres, a height of two decimetres, and opening in a thin plate against which the water rests, and is maintained, with its surface perfectly free, at a height of four decimetres above the lower edge of the opening.

"ART. 644. The right to a constant supply of water exists at every moment.

"ART. 645. The right to summer water (*aqua estiva*) exists from the equinox of spring to that of autumn; to winter water (*aqua jemale*) from the equinox of autumn to that of spring; and for water distributed at intervals of hours, days, weeks, months, or otherwise, for the time agreed upon or possessed.

"The distribution of water by days and nights is regulated by the natural day and night.

"The use of water on holidays is restricted to such holidays as were in legal existence at the time when the agreement was originally made, or actual possession of the water obtained.

"ART. 646. In distributions of water made by horary rotation, the time necessary for the water to flow to the outlet of an employer thereof shall be included in his period of rotation; and the water which passes down the common channel at the changes of the rotation belongs to the employer with whom the rotation terminates.

"ART. 647. The water which rises or leaks into the bed of the canal, subject to the distribution by rotation adverted to in the preceding article, cannot be stopped or appropriated by an employer, except at his own proper period of the rotation."

REMARKS ON PIEDMONTESE AND ITALIAN LEGISLATION.

Reviewing this legislation, we notice certain leading points bearing on each of the systems of delivery or distribution that have been mentioned.

FIRST SYSTEM—DISTRIBUTION BY VOLUME.

(*Sardinian Code, Articles* 643 *and* 641; *Italian Code, Articles* 622 *and* 620.)

The Sardinian Code.—First, with respect to the delivery of water by definite volume or quantity: The Sardinian code fixed a unit of measure for general adoption, which it called a *module*, and defined it as the quantity which would be measured out under certain simple conditions specified. And it provided that in all new grants or transactions concerning the delivery of waters by quantity, the amounts *should* thereafter be expressed in terms of this legal standard. (Art. 643.)

But the delivery of water in greater or less volume than the one module, for which the dimensions and character of orifice and head of pressure were given (Art. 643), of course necessitated the adoption of openings of different sizes, and the circumstances under which water was to be delivered, likewise made necessary, in different cases, its delivery under varied heads of pressure, and, hence, while the standard amount was fixed as a unit, the means of measuring out any number of such units or fractions thereof were left undetermined, and thus relegated to the field of hydraulic practice and that of agreement between the parties to the contract. (Art. 641.)

Providing for these cases, which, of course, really comprised nearly the whole practice, the law (article 641) left the choice of the form of the outlet of derivation to the parties to the agreement, but it held each to such choice, should the other insist, unless it could be shown that the resulting measurement was in error by an amount exceeding one eighth of the quantity contracted for. By this provision it was desired to prevent litigation over trifling amounts of water, and to promote care and insure greater accuracy in the preliminary determining of the conditions of measurement for each case.

This stipulation of a margin of one eighth for error, it was thought, was made advisable by the necessarily imperfect application of hydraulic rules, in the thousands of varying forms and dimensions of structures that circumstances would compel the use of. So the results of experience were called on for a guide, and the limit of an eighth of the desired amount of flow was held to be sufficient to include the variations likely to occur.

OPINIONS OF GIOVANETTI, DE BUFFON, AND SCLOPIS.

M. de Buffon, in commenting on this legislation, quoted from a Piedmontese writer, who is generally referred to as having been a high authority upon these matters, and I reproduce his remarks here:

De Buffon says: "M. Giovanetti, of Novare, a lawyer specially well versed in the questions which affect irrigation, has drawn up a learned work, in which he passes in review in a comparative manner all the Piedmontese legislation on this subject. In this work he expresses, on the subject of the module of water adopted in this country, an opinion similar to the two preceding. He makes amongst others the following observations:

"'In our article 643 (Sardinian code) there are indicated perfectly the conditions of a uniform supply; but in practice there are physical circumstances which rule, and it is necessary to content oneself with the least defective or the most practicable method. The essential was to establish a unit, to sanction a result without prescribing a fixed form. The law could not make prescriptions upon the form. They are in the domain of hydrometry, and can vary infinitely, either in accordance with local circumstances or in accordance with the progress of art.

"'Article 641 gives to the contracting parties the express liberty of making agreements upon the form of the orifice and the structure of derivation. These are they, then, who ought to settle accounts; and if the cultivator does not know how many cubic metres of water run through an opening of fixed dimensions, he knows very well what are the advantages which he can derive from this water in practice. The seller, on his part, also makes his calculations, and he bases them on the greater or less competition, and upon all the elements of the value of the water in a given locality. These reciprocal reflections determine the contract. An outlet is not constructed until an expert, in whom both parties have confidence, goes to the place and makes a report, submitted to the examination of the interested parties.'"—[De Buffon, Vol. II, p. 192, quoting Giovanetti.

De Buffon himself thought that a unit of measure should have been fixed at some universally recognized volume, and that the form and dimensions of the outlets for a considerable range of cases should have been determined by governmental action—by a ruling of the administrative department—and, hence, he criticised the Piedmontese law; but the reason for this shortcoming is well set forth by another author whom he quotes. The count Sclopis, member of the senate of Turin, who, in a memoir communicated to the Academy of moral and political Sciences, of Turin, said:

"'The Sardinian law has held a just middle course upon this point, by having due regard to ancient customs and acquired rights. In fixing the new unit with conditions, which, if they are exactly observed, correspond to a well determined discharge, it has deemed it convenient to confine them to the two most important: that is to say, the dimensions of the orifice, and the stipulation that the water must flow through it by simple pressure.

"'Apart from these two principal conditions, it has presented nothing, either on the form of the measuring apparatus, or on the nature of the precautions to be taken to maintain the constant and uniform pressure. In doing this, it has aimed not to tie itself to the results of generalized experiences, nor to fetter the progressive march in the application of hydraulic science with the ever varying circumstances of time and locality.'"—[De Buffon, Vol. II, p. 193, quoting Sclopis.

From this we see that regard for "ancient customs and acquired rights" stood in the way of what was really the right thing to do, in the establishment of a system of water measurement for the country; and we see an evidence of the difficulty of changing "ancient custom," even if it is wrong, in irrigation practice; and we note how far "regard for acquired rights" may influence or prevent legislation that it is supposed will affect such rights, even when, in truth, the effect would be to their benefit, if they were held on a just basis, and even when the proposed measures are in reality the right ones for all concerned. And we should learn, even from this small matter, how dangerous it is to allow rights to grow up, unregulated and unrecorded in intelligible form, for it will be seen later that Italy afterwards felt obliged to do in this matter what Sardinia failed to do, but what De Buffon thought should have been done years before. But, the Sardinian legislators, if not clear on the matter of water measurement, certainly understood the importance of having the extent of rights known as fully as possible, according to their unit of measure, such as it was.

IMPORTANCE OF SETTLED CONDITIONS.

Taking up the next point concerning this system of "Distribution by Volume":

Piedmont—The Sardinian Code.—Experience had taught the embarrassments which arise in the management of hydraulic works, in consequence of the long continued existence of questions open even to the extent of this small marginal limit. So, considering three years to be sufficient time in which to test the working of an outlet, the framers of this law inserted the clause barring all appeal from an agreement as to the form and dimensions of a measuring opening, when such three years from the time of commencing its use should have passed, even though the variation in the amount delivered, from the agreed upon quantity, exceed the one eighth limit (article 641).

But it was also well known that the discharges of such outlets were necessarily calculated for nearly fixed conditions in the canal of supply, and that any material variation in such conditions would produce a decided variation in these discharges, and hence it was provided that the right of appeal from an agreement concerning the form and dimensions of an opening to deliver any certain quantity of water, should hold good in all cases and for all times, should the normal flow or regime of the canal of supply at the point of the outlet be materially changed in any way (Art. 641). Here again the whole subject was thrown into the domain of hydrometric practice, and the courts had authority vested in them to appoint a hydraulic engineer to expert each such case which caused contention.

Lombardy.—The difficulty of dealing with acquired rights, and the way in which the best measures are put off because of the fixity of ancient customs in the practice of irrigation, is still further attested in connection with this subject of measurement of waters, by the Lombardian experience.

Articles 13 and 14 of the decree of 1806 treated the matter in this way:

"Until there has been established a uniform measuring apparatus and a common unit for the gauging of waters the construction of regulated outlets are continued to be made according to local usages."

"In provinces where no fixed measuring apparatus whatever is in use, the direction-general shall determine on one which must be in accordance with the local circumstances."

Now there had grown up in every one of half a dozen or more districts of this province a local system of measuring waters, and they were one and all defective, and the engineers of the country knew this and so represented it, but, local prejudice was such that, so long as Lombardy remained in this respect independent of the rest of Italy, these differences were never reconciled and there never was any definite unit of measure adopted, although the practice was gradually better understood as the engineers succeeded in introducing better forms of apparatus and experimenting in their results.

The Italian Code.—Attention is now asked to articles 622 and 620 of the new Italian code (see appendix II), which correspond to the articles 643 and 641 of its predecessor, the Sardinian code, upon which comments have been made.

It will be seen that in these two articles the new code for all Italy follows the old code for Piedmont, and other parts of northern Italy, in general principle, but differs materially in the expression of some of its details.

Article 622 of the Italian code, following 643 Sardinian code, prescribes a legal "module," and makes its adoption for the future obligatory in all agreements concerning waters; but instead of defining this module only as an amount of water which would pass out of a certain orifice, with a certain pressure, it gives it a fixed and determined volume for all cases, and says "it is a body of water which flows with the constant volume of 100 litres per second," and for subdivisions it may be "divided into tenths, hundredths, and thousands."

Thus after thirty years of trial the views of De Buffon, first expressed in the early part of the period, are shown to have been sound, by the Italian government doing in this particular substantially what he said the Sardinian government should have done.

Article 620 of the Italian code corresponds to article 641 of the Sardinian, and follows it quite closely in all but the clause concerning the limit of one eighth, allowable as an error in measurement before an appeal from an agreement might be taken.

This clause the Italian article leaves out altogether, thus testifying to the fact that hydraulic science had in the meantime advanced so much that there was no longer any reason for any such provision: if the parties to an agreement concerning the delivery of a stipulated volume of water had the orifice for its delivery properly calculated and adjusted according to the present state of the hydraulic science, it should be correct so long as the conditions of the canal and water supply remained the same, and there should be no margin allowed, no possibility of error forecast, and no appeal from the agreement; and if they did not have the calculations and construction properly made at first they ought to be made to stand by the agreement and structure as it was. Such appears to have been the reasoning of the framers of the new code.

SECOND SYSTEM—DISTRIBUTION BY USE OR SERVICE.

(*Sardinian Code, Article* 642; *Italian Code, Articles* 621, 653, 654.)

The provisions thus far spoken of were for cases wherein water was to be delivered under agreement in a certain volume through a specified outlet. The second and third paragraphs of article 641 made provision for settlement of disputes arising under agreements wherein no particular form of outlet was specified, which cases could only come up under agreements made before the passage of the law under consideration, for after its passage all contracts concerning water discharges had to be drawn in terms of the standard module, and the form and dimensions of the proposed outlets were required to be written in the agreement, else it would not be legal and binding on either party.

For these cases to come up under old agreements, in consideration of the embarrassments before mentioned as resulting from the long continuance of open questions as to measurement, yet to allow ample time for their adjustment and not to bar cases of recent development, the time in which an appeal from an agreement might be taken was fixed at ten years, with, however, the reservation of right to appeal at any time, for reasons heretofore given, should a variation of discharge be occasioned by a change in the canal of supply or in its flow of water.

And, finally, this article 641, and for this class of cases wherein no agreement should have been made as to the form of outlet, distinctly provided for judicial decisions to be based on the judgment of hydraulic engineers, nominated by agreement between the parties to the contest, or, if they could not agree, then wholly by the courts. The full significance of this provision becomes apparent when we know that, under other general provisions of law in Italy, the number of expert witnesses which may be summoned in a case is limited according to the character and importance of the case, and the judge has discretion to say how many such witnesses shall be admitted in all. Hence, for the cases under the above provision of the article 641, the judge having named the number of experts that might be called, the parties were allowed to agree upon them, and the judge was to be guided by their report as to matters of fact and scientific and practical deduction; but, should the parties to the contest fail to agree upon the experts, the court was given the power to appoint them.

This was a wise provision, growing out of a long range of experience in contests over hydraulic questions of the most complex kind. The effect was to raise the character and standing of experts in such matters. As we have seen, none but registered and proven engineers were permitted to practice the profession; and such provisions of law protected the profession from the debasing influences of the partisan rivalry between the litigants. The hydraulician was made the judge of the science and art in the case, and was not permitted to appear as the partisan witness of either the one or other party to the contest, as unfortunately is the case in our American system of experting. To conduct such cases there were lawyers trained in physical science and hydraulics, called *engineer advocates*. The engineers were (and still are in such matters in all Italy) the court referees and advisers for all scientific matters of fact and opinion.

The Sardinian Code.—Passing on to article 642, we find here certain provisions relative to the cases wherein water is delivered under the second arrangement—that according to its use or service, or the area to be irrigated.

The law provided for this case that when an agreement had been made for water to perform any certain service, as for instance the irrigation of any specified crop or determined area of land, and meaning for one irrigation, or more, or for one season or year, or more, as the case might be, then the contractor should be obliged to deliver that quantity of water necessary for the purpose. And, in case of contests coming up on this point, this quantity was in practice determined, according to the facts and the results of experiences in point, by the courts, upon the evidence of experts chosen by the parties to the suit, as might be allowed by the court, or by the court itself, if no amicable agreement could be arrived at.

But to avoid the precipitation of contests on this point before the courts, the law provided for an amicable adjustment of disputes by the fixing of an outlet such "that the grantee shall receive the volume sufficient for the service agreed upon, but nothing more."

After what has been said in commenting upon article 641, the second paragraph of article 642 does not call for remark. It is apparent also that all the provisions of article 642 were intended for cases that would arise under agreements made before the passage of the law, for, as we have seen after its passage, only the one form or arrangement for delivery—that according to actual volume expressed in terms of the standard module—was to be considered lawful.

The Italian Code.*—Article 621 of the Italian code corresponds to that (642) of the Sardinian last commented upon, and follows it closely in meaning, although not in wording, with one exception—

---
* See, Appendix II.

that the limit of time during which an agreement may be appealed from when the works for an outlet shall have been peaceably used as built, is reduced from ten to five years.

On this point of distribution by use or service rendered, the Italian code contains a very important provision not found in the old code for Piedmont. It is embodied in articles 653 and 654, and is to the effect that when an agreement has been entered into to furnish water to irrigate any certain area of land or any certain cultivation, or for any fixed purpose, with a stipulation that the drainage waters shall belong to the party delivering the supply, then the user of the water cannot change its use in a way to consume more or reduce the drainage waters in volume.

The reason for this rule is apparent when we know that, for instance, on the same area of land some crops would require and absorb twice as much water as others, the method of applying the water being in each case suited to the cultivation.

And, furthermore, the additional matter in the new code provides that the user of the water cannot divert and use again any portion of the drainage waters escaping from the place of his use of the water agreed for, on the plea of having increased the supply in any way. If he has obtained water under an agreement for any expressed purpose with the stipulation that the drainage water is to belong and be at the disposal of the owner or controller of the supply with whom he has contracted, then he must refrain from using the water for any purpose than the one named in the agreement, and must let all the drainage waters flow off as they naturally would, or as agreed upon, whether increased from any other source of supply or not.

This provision, holding a user of water to the strict letter of his agreement, has doubtless been made necessary in order to avoid contests wherein the facts for evidence—as to increase of supply and drainage from other causes or sources than the one agreed for—may be very obscure and difficult of substantiation one way or the other; and it has the effect of making the wording of agreements more explicit, in order to cover all contingencies of practice in each case.

The following articles (655 and 656) of the Italian code do not refer to waters furnished under agreements, but to those to which a right has been attached as a servitude, and, hence, they will be spoken of elsewhere.

THIRD SYSTEM—DISTRIBUTION BY TIME.

(*Sardinian Code, Articles* 644 *to* 647; *Italian Code, Articles* 623 *to* 626.)

The arrangements for delivery of water by actual quantity and by service to be rendered, having now been spoken of, we come to the provisions of the law, relating to the delivery of water by agreement as to *time*.

In northern Italy irrigation goes on the year round, and, in fact, the most copious irrigations are conducted through the winter, although at that season the sky is much overcast, rain, or sleet rather, frequently falls, snow is not uncommon, so that the upper part of the valley of the Po is covered for weeks at a time with a snow mantle a foot or more in depth, and standing waters are frozen to several inches in thickness.

The irrigation at this season of the year is that of meadows, or *marcite* fields as they are called, and it is a practice of a high order in the art. The object is to provide green food for the cows of the dairy farms with which the country abounds, and for which it is remarkable, supplying immense quantities of cheese and butter for export.

These meadows are formed after very exact methods, so that their surfaces are shaped into long narrow ridges parallel to each other, and of such longitudinal and transverse slopes that the waters applied through a ditch situated along the crest of each ridge, spread out from it laterally, flow in a thin sheet down each slope of the ridge over and through the grass there growing, and find exit longitudinally by way of drainage ditches lying between each two such irrigation ridges.

Thus in the coldest weather, with snow a foot deep elsewhere, the surfaces of these meadows are kept clear by the slowly moving film of water over them; the ground is prevented from freezing, and the grass kept green and growing, is cut from time to time and fed in stables to the cattle. The waters of springs, or *fontanili*, with which these plain lands abound, as elsewhere explained, being much warmer, are preferred to the canal waters for this purpose of winter irrigation, and command high prices at this season, as indeed do other waters as well, in some localities. Thus it is that there is a distinct practice following through a stated season, known as winter irrigation.

As in other countries, so, of course, in northern Italy, the season of ordinary summer irrigation is well marked by the climate and the requirements of the soils and the crops cultivated, and, thus, it comes about that waters are contracted for in certain streams of flow or amounts, as "summer waters" and as "winter waters," meaning for the seasons of summer and winter irrigation.

Again, irrigation is conducted day and night, the twenty-four hours around, summer and winter. Some persons contracting perhaps for the use of water only in the daytime, others for its flow at night; and hence the expressions "day water" and "night water."

And, in the way of explanation of the technology of the articles of the code which are to follow, finally, agreements are made for water to be distributed amongst consumers by certain hours of flow to each in rotation; and hence the expression "horary rotation."

Thus with respect to *time* there are arrangements (1) for summer waters, (2) for winter waters, (3) for day waters, (4) for night waters, and (5) for hourly waters; and to guide or prevent the contests which might arise under agreements for such waters, the Piedmontese (Sardinian) code contained the provisions found in the foregoing articles, 644, 645, 646, and 647.

The Sardinian Code.—And now, commenting upon these rules, we find in article 644, a declaration as to the rights, when an agreement had been entered into, to deliver any certain "flow of water" for any definite or indefinite period of time, to the effect that so long, and for every moment, as the time exists, the right to the flow existed: unless there had been a reservation in the agreement, whereby the flow might be checked at some time, the obligation to deliver continuously under this form of arrangement had been incurred, and this article recognized it.

Passing to article 645, concerning summer and winter waters, etc., after what has been said by way of introduction to these rules, no comment is necessary here; except to call attention to the fact that the law carefully defined the application of these terms, thereby removing much cause for misunderstanding of agreements.

Article 646 treated of a point which had given rise to much dispute—the question as to whom the tail end of a water supply belonged, when, after the stipulated hour for change had arrived, the stream was to be switched off on to the property of another user.

For instance, time is kept for distribution, by a time schedule, of a certain stream of water to different users of it in irrigation, at the head of a distributing ditch. The consumer has a right to it for two hours, or any other stated time, and the question is—to whom does the water belong which is in the ditch, on the way down, when the gate is closed at the head of the ditch, at the end of the time of the turn.

This flow is called by the Piedmontese the "tail of the water" (*coda dell' acqua*), and the article now under consideration said that it belonged to the consumer who last had the use of the stream.

On this point, Baird Smith says:

"Until this point was settled by the code, it was occasionally in dispute, whether the loss of time due to the passage of the water from the canal of supply to the distributing gates to the different employers, should be borne' by the proprietor of the canal or by the consumers; it is established as a general rule that, when the water passes below the outlet of the common channel, any loss of time that may arise shall affect the employers only, being borne by each in proportion to his distance from the head, or from the outlet of the field which precedes his in the order of the rotation."—[Smith, Vol. II, p. 288.

It has been elsewhere remarked, that much of the irrigated regions of the valley of the Po are underlaid with water-bearing strata of gravel, and that the cutting of canals through them often opens sources of additional water supply to these channels. It is not infrequently the case that such source yields a very material part of the volume carried by a canal, and that, hence, it has a considerably greater amount of water at a lower than at a higher part of its course; or, that being a distributing channel, when the water is shut off at its head, it still continued to have a flow derived from springs in its bottom.

The right to use these waters was a subject of contention—employers below claiming that they did not belong to the canal or ditch owners; so, as the result of these contentions, came decisions of courts, which were incorporated into the code in article 647: The waters rising in or leaking into a canal or ditch were held to belong to its owner; and employers below could not use them except at their proper hours and with the stream of distribution delivered to them.

The Italian Code.—Articles 623 to 626 (see appendix II) of the new Italian code correspond to those of the former Sardinian code, last commented upon, and follow them in principle and terms so closely that a comparison of details is not called for.

In closing some comments on article 647 of the Sardinian code, Baird Smith, writing in 1855, said:

"It is not uncommon in the irrigated districts of Piedmont and Lombardy for parties to make mutual interchanges of their periods of rotation. Special cases arise in which water is wanted at special times by individuals not possessed of the right to irrigate at such times. They, therefore, effect an exchange of period with other parties, to whom an arrangement of the kind may be convenient, and, though the law is doubtful on the point—some decisions being in favor of, and others against, this proceeding—there does not appear to be any valid objection to its use, if it be guarded by the provision, that the other employers of the water-course* shall sustain no serious damage by the manner in which it is carried into effect. An analo-

---

* Distributing ditches are called water-courses by the English writers on irrigation.

gous custom is common in India: the positions of outlets on watercourses, held in common, are often changed, and so long as other parties do not suffer by this, the interests of agriculture are certainly promoted by its being freely made use of."—[Smith, Vol. II, p. 290.

As may be inferred from the above, there was no provision in the Sardinian code, on this point, but now we find in the Italian code—the outcome of longer experience—an article which covers the case, as follows:

"ART. 627. In the same canals the users may vary or exchange their turns among themselves, provided such changes cause no injury to others."

Probably the author, Smith, was aware of decisions at the time of his investigation already made and upon which this provision of the later code was afterwards predicated; but, even so, it is an evidence of close study of his subject that the rule was afterwards enacted into law as nearly as possible in the words in which he said it should exist.

SECTION II.

THE RIGHTS OF IRRIGATORS.

(1) CONTINUANCE OF WATER SUPPLY.*

In Italy, as in irrigating countries generally, where there has been a clashing of interests between the owners of canals, or holders of great water privileges, and the irrigators to whom the waters were distributed, special points come up. Many water-rights were established there by grant and prescription, in times when from a troubled condition of society, no thought was had of future agricultural masses of people with interests to be protected. A water-right aristocracy grew up; the canal owners claimed the right of absolute property in the waters held by them, and undertook to do with it as they chose. If they could get higher rents for it in one section of country commanded by their canals than in another, which was occasionally the case, they claimed the right to discontinue the supplying of irrigators where water was cheap, after their annual or term agreement had been fulfilled, and of leading it to the lands of those who would pay more for it.

On the other hand, the irrigators claimed that they had expended their means and labor in the building of distributing works and preparation of lands to receive the waters, and being deprived of their

---

* See Smith, Vol. II, pp. 138-261; also, De Buffon, Vol. II, pp. 210-212, and elsewhere.

supply was equivalent to being debarred the use of their property. They claimed that these water-rights were not rights of property in the waters, as in the sense of ownership of land; that the grants were made for the good of the country and not for the exclusive benefit of the grantees.

In Lombardy, these questions came to a head from time to time in great struggles between immensely wealthy and powerful interests, but it was only during the last years of the last century and the first years of this, that they were well disposed of on principle.

The courts and senates rejected the claim of absolute ownership and ultimate right of control set up by the water-right grantees and canal men, and recognized the right of irrigators to the continued use of waters which they had for a considerable time had at their disposal, and to use which they had constructed distributing works and prepared their lands; and several of the local senates decided that so long as the irrigator paid the water-rates, he could not be deprived of the use of the waters, and that a change in water-rates had to be fixed by arbitrators, appointed by both parties at interest.

THE RIGHT IN PIEDMONT.

This was the law in Piedmont before the various petty governments were set aside in the early part of this century, in the consolidation of the kingdom of Sardinia. When the commission was forming the Sardinian code, an article carrying out this principle was embodied in its draft, and was agreed to by all of the local ruling interests to be conciliated but in the case of the senate of Genoa, a locality where irrigation was not practiced. Here it was insisted that the right sought to be established was subversive of the rights of property; so the article was stricken out of the draft of the code.

But it was subsequently held that the law had been established and recognized for all existing irrigations, so that these were protected notwithstanding the failure to incorporate the article in the code; and projectors of new irrigations have protected themselves by securing long term leases on waters before preparing their lands. The canal and water-right owners had apparently recognized the situation and dropped the conflict, for at the time Baird Smith wrote, any landed proprietor could "obtain a lease of a given quantity of water, either in perpetuity or for a specified term, on paying the current price for it."

THE RIGHT IN LOMBARDY.

In Lombardy, also, this question came up in a most aggravated form. The holders of water rights "acted on the principle that they had a right to do what they liked with their own, and were in the habit of suspending arbitrarily the supplies of water disposed of by them to other parties under subordinate grants, of increasing as they thought fit the prices to be paid, and, in a word, of pushing to its utmost limits the right of absolute property purchased by them from the State."

As the outcome of a long series of struggles over this point, the question was settled very much as already described for Piedmont: the water-right holders were restricted in the operation of their rights of propertyship in the waters, and compelled to distribute them amongst the irrigators according to ancient custom, notwithstanding the fact that in most cases of the older rights they held the water as an absolute property by virtue of purchase from government.

Baird Smith says, of this claim of the water-right owners to do as they chose, in Lombardy:

"But an agriculture founded on artificial irrigation cannot advance as it ought to, under such an arbitrary system; and so, in protecting the irrigating communities, there gradually grew up a right, which, being acknowledged by the legislative tribunals, modified the despotism of the government grantees. This right bears the name of the *diritto d'insistenza*, and assures to a province, or commune, or association of irrigators, or even to individuals, a legal claim to a continuance of such a supply of water as they may have enjoyed for long periods of time, and on the faith of possessing which they may have incurred heavy expenses. So long as the irrigating community pays the water-rent fixed by the grantee of the canal, it cannot be arbitrarily dispossessed of its supply; and in the event of the proprietor of the water desiring to change the rates of payment, this must be done through the medium of arbitrators duly nominated by both parties."—[Smith, Vol. II, p. 138.

It is as a result of this class of troubles that we find all agreements between those who have water to distribute, sell, or lease for irrigation, whether the government, private individuals, or great corporations, and those who use the waters, are made for long terms, the minimum, as a general rule, being nine years, and, for greater volumes, twenty or thirty years, and, not infrequently, for ninety or an hundred, or in perpetuity.

These contracts determine in detail the terms of the transaction, and are recorded and stamped, even if for an insignificant amount. Their form and provisions will be spoken of elsewhere.

(2) THE USE OF SPARE WATERS.*

We now come to another class of contests between those who held the water, and those who wanted to use it. The case we have just

---

* See Smith, Vol. II, pp. 257-260, and elsewhere; also, De Buffon, Vol. II, pp. 200-204, and elsewhere.

considered is one wherein water having been used in irrigation by certain employers of it, under leases or rents, at a determined rate for considerable periods of time, the owner of the canal of supply desires to raise the rent at the end of a lease, or lease the water to other customers, thereby leaving his former customer without a supply.

The present case is that in which an owner of a spring, or a canal of supply, not having use for all the water himself, refuses to sell it to any one at a fair rate, but insists upon wasting it. One would suppose that such cases, in a country where water is so valuable, would never occur, but there have been some remarkable instances of this kind of abuse, which are so instructive that I reproduce an account of one of them here, as given by De Buffon, upon the authority of count Cavour who was minister of finance of the Sardinian government. In his manuscript notes to De Buffon the count said:

"I have seen an example of each of the abuses that the new code has tried to prevent. Here is one of them:

"In 1832 the marquis of Saint G——, farmer,* of the canals of the Vercellais, having quarreled with the marquis Pal——, his neighbor, had persisted, during eight consecutive years, in throwing away into the Po, two streams of water that the marquis Pal—— offered to pay him 12,000 francs ($2,400) a year for. To satisfy a personal antipathy M. de Saint G—— consented to lose nearly 100,000 francs ($20,000), and to cause at the same time to the agriculture of his country a loss at least three times as great.

"The new code put an end to this deplorable state of affairs; but a sentence of the senate of Turin, founded on article 560 of it, was necessary in order to force M. Saint-G—— to have his revenue augmented by 12,000 francs a year.

"This same marquis of Saint-G——, wishing to coerce the community of F——, to subscribe to an engagement, which they thought oppressive, refused during two years to let run on the lands of this community the *colatures*§ of his vast domains, for which he was offered 6,000 francs per annum. He preferred to waste them into the Po.

"Marquises of Saint-G—— are rare, but as they are not impossible, the law does well in taking away from them the means of injuring people less rich and powerful."—[De Buffon, Vol. II, pp. 203-204, quoting Cavour.

The Sardinian Code.—The provisions of the code to which the count referred were contained in the article of the Sardinian code, which here follows:

"ART. 560. Every proprietor or possessor of water may make such use of the same for himself as may seem to him good, or he may dis-

---

* The system of farming out the revenues of and distribution of the water from government canals is explained elsewhere; and the abuses which have grown up under it have been shown. The present is a case in point, illustrating what has been said.
§ The waste waters from meadows and rice irrigations.

pose of it in favor of other parties, provided always that no title or prescription exists to the contrary; but after having used the water himself, he is not at liberty so to dispose of it as to cause it to be lost, to the injury of lands at lower levels, which might have benefited by it without causing any back-water, or injuries of other kinds to the higher employers. Whoever may desire to avail himself of the water referred to is bound to pay a fair price for it, whether the supply be derived from a spring existing in the upper estate, or from a stream introduced by special grant."

It will be seen that the provisions of this article meet the case quite fully, and, as a matter of fact, the contentions on points of this class were stopped by a few decisions of the higher judicial tribunals, under it.

Mr. Baird Smith, from a former edition of De Buffon's work, also quoted Cavour's account of the case of the Marquis of Saint G——, and, in concluding the topic, himself made the following remarks:

"I think few will dissent from M. de Cavour's conclusion; for if it is ever necessary that a man should not have full power to do what he likes with his own, or that the duties of property should be enforced equally with its rights, surely it is when the very sources of agricultural progress are concerned. I think, therefore, that the principle of requiring every proprietor of water to place it at the disposal of his neighbors on equitable terms, after his own wants have been fully supplied, is one of great importance in the legislation of irrigation, and well worthy of adoption by us in the East, where great canals are in progress." *—[Smith, Vol. II, p. 258.

The Italian Code.—In framing the new Italian code, article 545, the article (560) of the Sardinian was closely followed, so that irrigators have now the same consideration on this point, for all Italy, that those of Piedmont had twenty and more years ago. But, also following the framers of the old code, those who made the new refrained from inserting any clause corresponding to the ancient *diritto d'insistenza* of Piedmont and Lombardy—the right whereby any water company could be compelled, by judicial action, to continue the serving of its old customers, and prevented from conducting its waters to other customers, leaving users of water of long standing without any. As before remarked, the law on this point was set for existing irrigations by the action of the local senates in the last part of the last and the first part of this century; and after the adoption of the Sardinian code new irrigation agreements have always contained a clause protecting the irrigators for long periods, from possible withdrawal of their water supply.

---

* Mr. Baird Smith's report was written for the English East India Company, operating in India.

SECTION III.

OBLIGATIONS AND RIGHTS OF IRRIGATORS AND CANAL MEN.

(1) OBLIGATIONS CONCERNING WATER SUPPLY AND USE.

Experience teaches that the relations between those who command, for distribution, the water supply of an irrigation region, and those who receive and use it, cannot be too clearly understood. The scale of efficiency of canal works and of energy in their management is such that, at best, it is in practice exceeding hard, in any particular case not at an extreme, to say whether a management has been to blame or not for a failure of water supply. A canal manager may be so often a target of ungrounded fault-finding on the part of irrigators, that he is hardened to their complaints, and becomes careless of their interests; or he may be parsimonious in the business management of his property, his canal works become inefficient, or, not repaired in time for the season of rising waters, are damaged, and the irrigators suffer because of short supply resulting from his neglect or bad management.

On the other hand, a failure of water may be occasioned to irrigators, by reason of circumstances beyond the control of the contractor for the supply; the streams may not bring down their accustomed quantity, or the works may be damaged by unexpected and overwhelming floods, so as to cause delay in delivery of waters for irrigation; or third parties may maliciously or through neglect cause damage to works, or otherwise interrupt the water supply.

The questions growing out of the relations here spoken of, were found, in northern Italy, to specially demand the establishment of general guiding rules, and accordingly we find such provisions in the laws of the country, as seem to fully meet the more important points, for misunderstanding, likely to come up.

The Sardinian Code.—In Piedmont the Sardinian code was quite explicit on these relations.

The obligations of those who had water to distribute to customers, for irrigation, concerning their duties with respect to delivery of the supply engaged, the conditions whereunder they were not to be held responsible should there be a deficiency in the water delivered, the stipulation as to a rebate on the water-rate in the event of certain conditions being presented, the recourse of recovery for damages, and stipulation as to who should join in an action therefor, are fully and so clearly set forth in articles 664, 655 of this code that I present them without further remark:

"Art. 664. In default of special agreements the proprietor or other granter of water from a spring or canal is under obligations to those who hold grants under him, to execute all the ordinary and extraordinary works required to procure the supply; to conduct and to preserve the water up to the points at which the employers take possession thereof; to maintain the structures in an efficient state; to repair the bed and banks of the spring or canal; to effect the usual clearances; and to exercise due diligence, watchfulness, and care to insure the delivery of the water, and its regular supply at the appropriate times, under pain of having to pay compensation for all injuries inflicted on the employers of the water by his neglect of duty.

"Art. 665. If, however, the granter of the water can prove that the deficiency of the supply arose from natural causes, or from the acts of others, for which he could not be held responsible, either directly or indirectly, he shall not, in such cases, be bound to pay compensation for the injuries sustained by the users of the water, but only to submit to a proportional diminution of the amount of water rent, or the equivalent corresponding thereto, whether previously paid or not, without prejudice to the right of the injured parties to institute an action for compensation against those who may have caused the deficiency.

"In the second cases contemplated above the granter of the water is bound to join in the action with the employers, should they so desire, and to use every means in his power to assist the same in obtaining compensation from those who had caused the deficiency of water."

The Italian Code.—Articles 649 and 650 of the Italian code (see appendix II) correspond to the foregoing, numbers 664 and 665 of the old Sardinian code, and are the same with one important exception.

The penalty which was to be imposed upon the contractor, to deliver water for non-fulfillment of agreement, as embodied at the end of article 664 of the Sardinian code, is not reproduced in the Italian. It may be, however, that other general laws of the country regarding contracts, or agreements, or other cognate matters in principle, amply cover the case, and enable the employer of the water to recover compensation for damages from the contractor to deliver it, should injury result from his neglect.

In this connection, although noticed before, under the heading of "Distribution by use or service rendered," articles 653 and 654 of the Italian code are worthy of mention, as imposing an obligation on the user of water not to change in any way its use, so as to affect the volume required, or the amount of drainage waters left over when the water is furnished under an agreement to do a certain service, and with a clause reserving the right to the drainage waters, even though a plea is advanced that the volume of the drainage waters has been increased from some other source.

Thus, water being furnished to irrigate a certain tract of land in a certain crop, under an agreement whereby the drainage waters were reserved by the party furnishing the supply, even though the irrigator should introduce a new and additional supply on to an adjacent tract, lying higher, and thereby increase the amount of drainage from the lower estate, he cannot use any part of said drainage, but must ~~let~~ it flow for the benefit of the party to whose benefit the reservation in the agreement has been made.

The apparent reason for this rule, being given in another place, will not be repeated here. It would appear, however, that nothing could prevent an agreement being made whereunder an irrigator would be fully protected in the use of any addition which he might cause to the drainage waters.

(2) PRIORITY OF PRIVILEGE IN DISTRIBUTION.

Principles strenuously contended for and contested, at one time or another in all irrigation regions, are those of priority of rights to water: first, by virtue of commanding localities on streams; second, because of antedate of claim; and third, because of contemporaneous advantage or stated claim for a definite time.

In the legislation of northern Italy these principles not only found recognition in the adjustment of rights to water from natural streams, but in the arrangement of the generalities of distribution from canals to consumers. This last was a feature peculiar to Piedmontese legislation, and found place in the Sardinian code, in articles 666 and 667, some points of which are worthy of explanation and remark.

Generally, the management of irrigation is such in Italy that, whether water is delivered by volume, as per module, or according to the use or duty assigned, or indeed, in all cases except when a continuous stream has been contracted for, the periods for each delivery are determined and adjusted in a schedule long beforehand, perhaps at the beginning of the season, on each canal. The same outlets, the same ditches, may, at different parts of the year, different months, different days, or different hours, serve different people, but it is known and recorded long beforehand at what times each is to receive his supply.

This being the case, and the schedule being determined, should the water supply be all engaged, and should, at any time, from any cause, a deficiency occur, as by the breaking of a canal or temporary derangement of any work, the loss of water had to be borne by the parties whose turn it was to receive it, according to the schedule. They did not have to pay for water which they did not get, but there being no water, or a short supply, during the hours, days, or weeks, as the case may be, for which their turn was set, they were the sole losers so far as the effect of slack supply was concerned. They could not be served with water at the expense to the shares of other consumers who had been booked for other hours, or receive water at any other time, unless there was a surplus of supply over the demand, at some period, which could be turned to them. (See article 666.)

In cases wherein water was not distributed by turns, but in a continuous stream, another rule prevailed. First, the principle of "first in time first in right" was applied: he whose engagement or agreement for water was the oldest, received his full supply so long as water lasted, while those whose contracts had been made more recently had to suffer loss by the deficiency; and so on down the scale as to time—the last one being the first sufferer.

Then came in the principle of advantage by reason of situation. Where privileges were of even date in origin, the one located highest on the canal of supply had the advantage to the full extent of its quota, while the ones below, commencing with the furthest from the head, had first to suffer reduction when the supply was short. But, as will be noticed in the law, no one was expected to pay for water which he did not get, and, if payment had been made in advance, the irrigator had a right to reclamation for the amount. (See article 667.)

These articles are now themselves transcribed for closer study:

"ART. 666. The deficiency of water shall be borne by those parties during whose period of rotation the said deficiency may occur; saving their right to compensation for injuries, to diminution of water-rent, or its equivalent as above defined.

"ART. 667. Among the different employers, those individuals whose titles or rights of possession are most recent, shall first bear the effects of the deficiency of the supply. Among employers equal in the preceding respects, the deficiency shall first affect those whose outlets are at the lowest levels; saving, in all cases, the right of action for compensation against the parties causing the deficiency."

The Italian Code.—Articles 651 and 652 of the Italian code* correspond with the foregoing, numbers 666 and 667 of the Sardinian code, and closely follow them with one exception.

The second clause of article 652 differs from that of 667, in that the expression does not clearly indicate that the principle of priority by reason of situation on a stream is to be applied. [Refer to remarks under article 652, in appendix II.]

---

* See, Appendix II.

AUTHORITIES FOR CHAPTER XI.

In the preparation of this chapter I have consulted and compared the following named authorities:

De Buffon.—[Work cited as an authority for Chapter IX. (French.)] See, Vol. II, B. VII, Ch. 38, Div. II, and Div. IV; Ch. 39, Div. II; Ch. 40, Div. I; and elsewhere as cited.

Smith.—[Work cited as an authority for Chapter IX.] See, Vol. II, P. III, Ch. I, Secs. I, II, III, IV, and V, and Ch. II; P. IV, Ch. I, Sec. III, and Ch. II, Sec. III; and elsewhere as cited.

Sardinian Code.—[Work cited as an authority for Chapter X.] See, Articles 560, 641, 642, 643, 644, 645, 646, 647, 664, 665, 666, and 667.

Italian Code.—[Work cited as an authority for Chapter X.] See, Articles 545, 620, 621, 622, 623, 624, 625, 626, 627, 649, 650, 651, 652, 653, and 654; and also the annotations to each of these articles.

CHAPTER XII—ITALY[4];

REGULATION OF DRAINAGE AND WORKS CONNECTED WITH IRRIGATION PRACTICE.

> SECTION I.—*Regulation of Works.*
> Construction of Works on Private Lands.
> Distances from Boundaries of Estates.
> Construction and Maintenance of Works—Free Passage.
>
> SECTION II.—*Rights and Obligations of Drainage.*
> Drainage Complications.
> Principles of the Italian Laws.
> Provisions of the Codes—Sardinian; Italian.
>
> SECTION III.—*Sanitary Legislation.*
> The Unheeded Teachings of Experience.
> Sanitary Effect of Unregulated Irrigation.
> Regulation of Rice and Meadow Culture.
> Sanitary Regulations—Modern Legislation.

SECTION I.

REGULATION OF WORKS ACCESSORY TO IRRIGATION PRACTICE.

DISTANCES TO BE PRESERVED FROM BOUNDARIES.

In irrigation regions closely settled and fully developed, questions frequently come before the courts which are rarely, if ever, met with in other countries, and thus arise necessities for provisions of statutory law which would be altogether needless elsewhere. Prominent amongst these questions are those relating to the rights of individuals to do as they please with their own property or on their own lands.

The class of operations accompanying or forming a part of irrigation practice, are peculiarly of a character whose effects are not and cannot always be confined to the possessions of those who carry them out. Indeed, the more important works must necessarily be community works; in nearly all there is a community of interest or a widespread effect, and even works for one's own benefit, solely, on one's own property, not infrequently infringe upon the rights of one's neighbors to an extent that it becomes necessary to impose restraint on the acts of individuals in exercising their rights of propertyship.

Instances of this class of legislation have already been cited in former chapters of this report; notably under the headings concerning springs, and water-rights, wherein the right to excavate, bore, or dig for water on one's own property is limited by law out of consideration for the rights of others having springs, wells, or water supply works on adjoining lands.

We now come to certain provisions of law which limit the right of individuals to construct canals and ditches for conducting waters, on their own property. A moment's consideration shows the necessity for the restrictions.

The reasons may be summarized as follows:

When canals or sources of supply are situated, as often they necessarily are, near the boundaries of the tract on which they lie (and which perhaps may be a narrow strip condemned, on which to construct the work only), by reason of the percolative nature of soils or subsoils, if parties owning the adjoining lands were allowed to excavate a parallel work as close to the border of their lands as they chose, the waters of the canal or source adjoining might thus be caused to percolate away into the new excavation, perhaps at a lower level, to the great injury of him or they who own the source which has produced, or canal which has brought them.

Or, by reason of the instability of the soil itself, if persons were permitted to excavate as closely as they chose to a boundary of their lands, the ground itself might be caused to cave away from a canal bed or bank, fountain, basin, in adjoining lands to the great loss of its owner.

And, again, canals used for carrying waters, liable to the erosive effects of their currents, if constructed close to the bounds of one estate, cut in upon the lands of others, to their injury; or, being in porous soil, impart undesired moisture to the lands of others, thus rendering them unfit for cultivation.

As a consequence of experiences of such effects, we find in the laws of northern Italy a number of provisions intended to meet the cases to which they give rise, or to prevent the cause of such cases. Amongst these are the following articles, of the Sardinian code formerly ruling in Piedmont, which, after this introduction, so far as our subject goes, require no further explanation:

ARTICLES OF THE SARDINIAN CODE.

"ART. 599. The ditches and canals which the proprietor of an estate may excavate on his own land, shall be placed at a distance from the boundary lines of adjoining estates at least equal to their respective depths, except in cases where local regulations prescribe a greater distance.

"ART. 600. The foresaid distance shall be measured from the edge of the bank of the ditches or canals nearest the boundary lines above referred to. This bank must always have a slope equal to its height, or, in the absence of such a slope, it ought to be provided with retaining works.

"Where the boundary of an estate is formed by a ditch possessed in common, or by a private road also common, or subject to the servitude of passage, the distance shall be measured from the crest of the bank, as above defined, to the edges either of the common ditch or road, nearest to the property of the party desirous of excavating the new canal or ditch; the obligations regarding the slope or revetment of the channel remaining in full force.

"ART. 601. Should the ditch or canal be excavated in the vicinity of a wall possessed in common, the observance of the foregoing distance is not necessary, but the party excavating the said ditch or canal shall be bound to construct all such intermediate works as may be necessary for the protection of the wall."

It will be noticed that these rulings simply prescribe minimum distances to be observed in the location of works. In the judgment of the courts as advised by professional experts, in each case a greater distance might have been insisted upon, or other precautions enforced, as was indeed frequently the outcome in practice.

*The Italian Code.**—Articles 575, 576, and 577, of the Italian code, follow closely the foregoing numbers 599, 600, and 601, of the old Sardinian code in wording as well as meaning; there being only a slight difference in the framing of the second one of the three, which really does not materially change the meaning, except that the degree of slope required to the bank of a ditch when adjoining a property line is not determined in the new law, as it was in the old; thus leaving this matter of detail to administrative regulation or judicial decision.

OBLIGATION AS TO CONSTRUCTION AND MAINTENANCE OF WORK.

A very important ruling, in the way of a regulation for the construction and maintenance of canal works, was embodied in the Piedmontese law, in the form of an obligation upon the owner of any ditch, canal, or water-course, to so plan, lay out, construct, and maintain it, that neither the work itself nor the flow of its water, should interfere with the free passage of travel on public or private roads or paths, nor with the free flow of waters in and efficiency of other canals or ditches, whether for irrigation or drainage purposes.

The Sardinian Code.—This provision was made in article 633 of the Sardinian code, which was as follows:

---
* See, Appendix II.

"ART. 633. In cases where waters flowing for the benefit of individuals prevent the adjoining proprietors from passing freely to their estates, or check the circulation of water in other irrigation or drainage lines, the parties benefiting by the waters are bound to construct and maintain in good order the bridges necessary for intercommunication, in a sure and convenient manner. They are farther bound to construct and maintain such culverts, aqueducts, and other like works, as are required for the free progress of irrigation or drainage, saving an agreement or legitimate title to the contrary."

The Italian Code.—In framing the new code for Italy, the old Sardinian ruling was closely followed in this particular, as will be seen from article 608 of the Italian code.*

SECTION II.

THE RIGHTS AND OBLIGATIONS OF DRAINAGE.

DRAINAGE TROUBLES IN ITALY.

More than in France, Spain, or any other country where irrigation has been broadly practiced, in northern Italy the problems of drainage have been ever present in the practical, legislative, and administrative complications which have been developed by it. This has been quite naturally brought about: for irrigated northern Italy is a well watered country, both with respect to the number and volume of streams, which course from the adjoining mountains across its plains, sometimes producing widespread and disastrous floods even in the irrigated districts, and also in the amount and regularity of its rainfall. And not only is the country thus well supplied with water by the streams from the mountains, and directly by the rains from the sky, but its subsoils abound with flowing waters which break forth in many living springs thickly scattered over wide regions of its plains.

Irrigation without drainage, and systematic and thorough drainage too, in northern Italy, would very soon result over the whole country in disaster: financial failure in agriculture, and a general depopulation of the country, because of its unhealthfulness. Drainage, then, is not only an essential to individual success in irrigated cultivations, but it is a requisite to the maintenance of the health of populations, and hence we find that it has received very close attention not only as an art, but as a social and political problem.

And still again, in most cases, the waters of drainage are not carried to waste—they are property, valuable for the irrigation of other

---
* See, Appendix II.

lands, and almost as much in demand for the purpose as are those of springs or rivers.

Authorities unite on this point, but, for present illustration, I cite only one. Baird Smith says:

"As the necessary complement to an effective system of irrigation, arrangements for disposing of the drainage-waters connected with it are essential. It will, I believe, be found in most cases, and I know from experience it is especially so in northern India, that imperfections of local drainage, as connected not so much with the great topographical features of the country, as simply with irrigation itself, within the limited area it affects, are more frequently the source of malaria and injury to the land than anything else."—[Smith, Vol. II, p. 300.

In another place, speaking of troubles in the irrigation regions of India, this same author has written:

"A comprehensive and authoritative system of drainage in connection with irrigation must be matured, and duly sanctioned by government, before the existing evils can be wholly eradicated."—[Smith, Vol. II, p. 303.

PRINCIPLES OF THE LAW.

The Sardinian Code.—The civil code which ruled in Piedmont contained some very important provisions relative to rights and obligations connected with drainage, yet there was much by way of regulation of private and public works, and the acts of individuals, governing and affecting drainage matters, that was left to the discretion of the administrative authorities in the execution of their general police power.

We find in the code express provisions on the following points: (*a*) The right of natural drainage-way, which assured to the owner of lands the continuation of the privilege to have waters, draining naturally from his estate, flow off on to lands below, as they by nature were accustomed to flow, even though they injured the properties lying there. But this right was accompanied by a stipulation that such waters were not to be increased artificially; and at the same time by a prohibition upon the owner of the lower lands not to interfere with their flow. (See Art. 551.)

(*b*) An obligation upon the proprietors of lands where were situated channels or embankments serving for or necessary to the preservation of efficient drainage, to keep such works in repair. (*c*) An obligation on these proprietors to construct new or additional works, such as might be necessary for the preservation or protection of the existing structures, channels, or banks. (*d*) An obligation on all land proprietors to keep the water-courses and channels through their estates clear from such deposits or accumulations of material as might interfere with the free escape of the waters, and thereby cause damage to their neighbors. (*e*) The right of interested parties, suffering or threatened in their estates with damage by reason of the necessity for repairs to works, the removal of deposits or clearance of channels, situated on others' lands, to go on to such lands, and themselves make the repairs, removals, or clearances. (*f*) The right of the party thus in peril by reason of the necessity for additional protective or other works, to go on the lands of another and there construct them. (See Arts. 552, 553.)

(*g*) An obligation upon all land proprietors interested in the maintenance of channels and embankments, or the preservation of free escape for drainage waters, to contribute in proportion to the extent of their interests towards the expense of such maintenance and clearances.

(*h*) The individual right of proprietors interested in such works and channels to proceed for the recovery of damage resulting to them by injury to the works, or by obstructions made or caused to form in the channels, against the party, or parties, causing such injury or formation. (See article 554.)

But the rights (*e*) and (*f*) were accompanied by the stipulations:

That, in exercising them, the property of others was not to be injured.

That, before exercising them, special authority should be had from competent local administration, or judicial officers. And that all interested parties be heard by such officer before the authorization should be issued.

And the declaration of these rights was also accompanied by that of an obligation, on the part of parties desiring to exercise them, to conform to administrative regulations applying to the water-courses, or other channels, or works of the locality. (See articles 552 and 553.)

PROVISIONS OF THE CODES.

Sardinian Code.—The foregoing points were embodied in the code which ruled in Piedmont, in the articles here presented for reference:

"ART. 551. Lower lands are subject towards those which are higher to receive all the waters which flow naturally, and without the aid of artificial works, from such higher lands.

"The proprietor of the lower estate shall not raise any embankment whereby this escape may be interfered with.

"The proprietor of the upper estate shall refrain from doing anything whereby the servitude of the lower land may be aggravated.

"ART. 552. When the channels or embankments which serve to contain waters within an estate are broken down or destroyed, or when variations in the course of the water render defensive works necessary, and the proprietor of the estate fails to restore the channels and embankments, or to construct the required works, then those who shall suffer injury, or shall be in imminent danger of it, can cause the works to be executed at their own expense; they can avail themselves of this power, however, only on the condition that the proprietor of the land on which the works are to be constructed shall suffer no damage; they must, furthermore, receive beforehand the permission of the competent authority, to be given after the parties interested are all heard; and also must conform in all cases to any special regulations which affect the management of the waters.

"ART. 553. The same rule shall apply when it is considered desirable to destroy or remove any obstacle to the free escape of waters in the form of deposits or collections of other materials, within an estate, or in a private water-course, the existence of which threatens injury to adjoining lands.

"ART. 554. All the proprietors who have an interest in maintaining the channel and embankments, or removing the obstacles referred to in the preceding articles, may be called upon to bear their shares of the expense incurred, which shall be rated on each in proportion to the benefit he receives from the works. In every case the proprietors shall have the power of proceeding individually against the party or parties who may have caused the destruction or choking up of the channels referred to, for the amount of the expenses incurred, and for compensation for damages caused."

In addition to the preceding clauses expressly pertaining to the subject of the natural right of drainage, and the obligations or rights of proprietors, relevant to the maintenance of drainage works and channel-ways, the Sardinian code made provision for the acquirement on the part of individual land proprietors, of the right, as a servitude, of conducting drainage waters, across the properties of others, and for the acquirement, by condemnation, on the part of works declared to be of public utility, of titles to the necessary strips of lands for purposes of construction of drainage works of all classes.

These provisions and others relating to rights of way for drainage will be noticed in the chapter about rights of way for waters, generally.

Another branch of this class of legislation is the regulations relating particularly to sanitary matters, and which will also be noticed in the next section of this chapter.

The Law of Lombardy.—This subject of drainage was of great importance in Lombardy as well as in Piedmont, and commanded special attention, also, in the laws of that country, of which some evidence will be found in the third division of the general law of that country, concerning irrigation associations, transcribed in the chapter which follows this.

The Italian Code.—The new code for all Italy, superseding the laws for Piedmont and for Lombardy, makes equally full provision for the interest of drainage. Articles 536 to 539*, inclusive, contain matter to almost exactly the same effect as the articles 551 to 554 of the Sardinian code, already analysed.

The subject of right of way for drainage waters, as will be seen hereinafter, in the chapter on rights of way, is also, quite as fully considered.

SECTION III.

SANITARY LEGISLATION.

THE UNHEEDED TEACHINGS OF EXPERIENCE.

No branch of legislation affecting irrigation interests, under the several governments of northern Italy, has been more often the field of enactment than that having for its object the preservation or promotion of good sanitary conditions in irrigation regions; nor has any other class of irrigation legislation been subject to such frequent fluctuations and amendments, or to such radical changes.

The necessity for this class of legislation has been the result of the gradual development of irrigation without proper system in the arrangement of works and without due care in the management of the waters and cultivations: it has been the natural outcome of a practice wherein every individual has striven for his own special advantage, and no consistently and constantly-acting overseeing power has cared for the interests of all, by guiding or controlling a little the actions of each.

To the reader of the annals and the observer of the development of irrigation, reclamation, and drainage practices, it cannot but seem that no people ever would or ever will profit by the former experiences of others in the lines of their intended endeavors. In our day and country one daily sees or hears of projects, theories, or practices, being put forward, the like of which have elsewhere long ago been tried and proven unprofitable, inadequate, or harmful. Personal experience or observation seems to be the only teacher in these lines of knowledge, which those who embark in hydraulic agricultural enterprise admit to their counsel. Professional knowledge of or technical data concerning what has been done elsewhere, with its results and lessons, it would appear had no existence. Superficial observation and blind experiment, guided by the illimitable self assertion of the times, which a plethoric purse prompts or narrower

---
* See, Appendix II.

views of self interest stimulate, guide some of our most important enterprises; others are reined by those who look to immediate self aggrandisement, without reference to the legitimate outcome at all; still others, by those who do not know but that the field of their experience is almost a virgin one, and that irrigation, for instance, is a Californian invention.

Seeing that these great interests are here developing under such influences, what must have been the surroundings of the growth of irrigation in Italy several centuries ago, we may well imagine. That the circumstances were unfavorable to the realization of the best results, we may well understand; that the results in many respects were very bad, there is ample evidence at hand.

SANITARY EFFECT OF UNREGULATED IRRIGATION.*

Not only is all irrigation, where conducted without adequate natural or artificial drainage of the soil, and as ordinarily practiced to effect anything like a full development of the capabilities of lands, harmful and injurious to the healthfulness of the irrigators and residents of the region irrigated, but certain cultivations in themselves are unhealthful, and necessitate the use of waters in a manner which produces an unsanitary condition of their neighborhood.

Trouble of this character made itself apparent in Italy during the fifteenth century, and following upon the introduction of rice cultivation which had been brought into the Venetian provinces in the early part of the century, from Spain. Now, the experience in Spain should have constituted a lesson for the Italians, but it did not. Rice had been brought into Spain by the Moors full two centuries before, and its cultivation had been the cause of most serious fever epidemics and such widespread alarm that regulative measures had been enacted from time to time, and at other times the cultivation had been prohibited altogether by royal decree, and then again allowed under stringent rulings as to locality and the provision of proper drainage.

In general terms, this same experience has been repeated in Italy. The cultivation of rice was first introduced upon marshy tracts unsuited for other cultivation without expensive reclamation and drainage, and at localities somewhat remote from thickly settled districts. It then spread, by degrees, into the lands irrigated from the great canals, and in the best neighborhoods of the country, approaching the gates of the large cities and the villas of the upper classes of society.

---

* Marsh: Rept. Dept. Agri. 1874, p. 366; Smith, Vol. II, pp. 219–224, 319–328, and elsewhere in Vols. I and II; De Buffon, Vol. II, pp. 151–161, and 339, *et seq.*

Fever epidemics became more and still more prevalent, and many cases of low fever were always present, even when not epidemic, so that from time to time there arose most violent opposition to rice cultivation at all, and there was a constant demand for its regulation.

LEGISLATIVE REGULATION OF RICE CULTURE.

Lombardy.—In Lombardy the earliest sanitary regulation of which there is record was promulgated in 1575. It took the form of restricting rice cultivation to certain areas, and prohibited it within certain distances of inhabited places. From that time on to the beginning of this century the records bristle with regulations promulgated, modified, annulled, and reënacted. In the territory of Milan, for instance, in 1583 the cultivation of rice was absolutely prohibited. In 1593 this was modified by a regulation forbidding rice cultivation within six miles of the city of Milan and within five miles around every other town; and at later dates these distances were successively increased and diminished as the rice cultivators found favor by fair means or foul with the rulers, and as the healthfulness of the country permitted popular sentiment to cool off on the subject, or the unhealthfulness roused the people to vigilance again.

In 1630 a frightful pestilence swept over the province of Milan. Rice cultivation was again prohibited for a short time, but again became prevalent. At a later date the distances from the cities, within which rice might be cultivated, were reduced from "long" miles to the same number of "short" miles and these were to be measured from the centers of the towns and not from the ramparts. And so matters ran on until the beginning of the present century, when Napoleon, formulating the experience of the past and calling to his council the best informed people of the country, promulgated the regulation which remained as the law of the land at least up to the time of consolidation of the present government of all Italy, and then become the foundation for the newer and present rulings.

Piedmont.—In Piedmont the sanitary regulation of irrigation was first seriously attempted in 1608, when the cultivation of rice was prohibited in any part of the kingdom, except by royal special permit; and it was stipulated that lands to be used for rice cultivation should be confined to those unfit for producing any other crop, and should be situated at least about four and one half English miles from any town or village, and six hundred and fifty yards from any road; that the consent of the heads of two thirds of the families in the commune should be obtained in each instance, and there should be an obligation on the part of the holder of the permit to secure and maintain perfect drainage for his fields, under the supervision of the government engineers. There were heavy fines named for the violators of this law, and other provisions made for its enforcement.

This species of culture had already grown to considerable magnitude in certain parts of the country, and there was much capital interested in the lands and canals devoted to it, consequently there was a perfect storm of opposition to the law. The chronicler, hereafter to be named, says that "complaints *rained down*" upon the government authorities, so that in 1663 the order was modified, so as to prohibit the cultivation of rice within four and one half miles of Turin, three miles around Vercelli, nine hundred yards from other towns, and seventy-five yards from the roads. Then, in 1667, the cultivation was absolutely forbidden in certain parts of the country. And thus the history goes with alternate prohibitions, limitations, regulations, and licenses from that time down to the year 1855, when a commission or committee of the senate of Sardinia was appointed to inquire into and report on the whole matter.

This committee reported a history of the legislation of the subject, from which the foregoing brief recital has been drawn, and it then expressed its opinion and made its recommendations, in language substantially as follows:

"Three conclusions appear to us to be deducible from the rapid review just given of the laws affecting rice cultivation, which have grown up among us during the course of two centuries and a half.

"First, that the sole remedy against the insalubrity of rice irrigation, which has been applied in practice, has been to keep it at a distance from inhabited places; but that the limit of this distance has been increased or diminished in a manner wholly arbitrary, and without reference to any theoretical principles or experimental results which warrant the terms selected. We say this was the sole remedy, because, although the laws ordain that free passage shall always be insured for the water, no specific plans for drainage were either suggested or enforced; and the districts where rice cultivation prevails, remain still unprovided with this important means of securing their salubrity.

"The second inference which appears to the committee, no less than the first, is, that a remedy which has been altered incessantly, and at brief intervals, cannot be regarded as a successful one, since it must have failed to produce the results anticipated from it by those who tried it in the various forms.

"Thirdly, it is clear that throughout the entire progress of our legislation it has always been found necessary in endeavoring to limit the extension of rice irrigation to respect the interests which have grown up in spite of the laws, and to sanction the continuance of the culture in places where it had been established for considerable periods.

* * * * * * * * * * * *

"The discontents and difficulties created have been such as invariably to force the government to modify its orders and to admit so many exceptions, as, in point of fact, rendered the laws nearly inoperative.

"If, therefore, the ancient laws do not supply examples of successful remedies which we can imitate; if, further, the facts on which a definite law could be founded so as to secure the confidence and respect of all parties concerned do not at this present moment exist, the committee is of opinion that measures should be taken to collect such facts, and that all attempts at final legislation should be deferred until this preliminary inquiry has been satisfactorily completed.

"On the other hand, the committee is distinctly of the opinion that certain conditions should be attached to permissions to form new rice lands; and, pending the collection of facts on which a final law may be based, they think that a temporary measure may properly be sanctioned. They therefore recommend that the project now submitted be passed by the senate, with the modifications which have been suggested by the committee."—[Smith, Vol. II, p. 326.

The measures recommended by the commission were enacted into law, and remained as the law of the country, at least until it was merged into the present kingdom of Italy.

The chief points in this law will be given under the next subheading of this chapter.

SANITARY REGULATIONS—MODERN LEGISLATION.[*]

As I have before remarked, the necessity for regulation of irrigation, because of sanitary reasons, did not apply only to rice cultivation, although these great contentions and oppositions have come up over attempts to prohibit or put a limit on the extension or continuance of the irrigation of this crop.

The modern regulations providing for the preservation of sanitary conditions in Lombardy, specially applied to all meadow cultivations by irrigation, as well as to the fields devoted to rice raising. I here present an abstract of their principal points, and then pass on to the laws proposed by the committee and voted by the senate of Sardinia, for Piedmont.

Lombardy.—The following is an abstract of the principal points of the irrigation sanitary regulation for Lombardy—promulgated under a law of 1809.

The establishment of new rice fields without special permission of the prefect of the department, was prohibited under pain of a heavy fine upon both the owner of the land and the tenant.

Permits were granted for such establishments only on lands situated at least five miles from the capital of the kingdom, three miles

---
[*] See, Smith, Vol. II, pp. 225–231, and 328–331; also, De Buffon.

from towns of the first class and fortified places, and one and a quarter miles from towns of the second class, and five hundred and fifty yards from the smallest towns; and these distances were to be measured at right angles from the exterior limits of the towns.

Cultivations of rice already existing within the limits specified from the capital were to cease within three years after the promulgation of the decree, and the lands be cultivated in other crops; under pain of a heavy penalty. Those existing within the limits prescribed from other places, were to be subject to further regulation after due inquiry in the communes where situated.

All rice cultivations were to be conducted in accordance, as to drainage, with regulations prescribed for each case.

The establishment of meadows, whether constantly or periodically irrigated, was prohibited within the limits of thickly inhabited places, and all such meadows were ordered abolished and the cultivation changed before the expiration of the then present year.

Permits for the establishment of meadows were to be granted only for lands situated at least eleven hundred yards from the walls of the capital city, and five hundred and fifty yards from those of other places; and in accordance with plans which were intended to insure proper drainage of the fields and disposal of the drainage waters.

Other regulations dated in 1817, prescribed the forms necessary to be observed in applying for and obtaining these permits—amongst which were the submission of plans of the fields to be laid out, examination of them and of the grounds by the government engineers and local authorities, publications of intention, hearings of opposition, reports of engineers and local officers, etc.

Piedmont.—The law reported by the committee, and passed by the Sardinian senate in 1855, and of which mention has been heretofore made, contained points as follows:

A registration of rice cultivations was to be enforced, and heavy penalties were prescribed for the establishment or continuance of rice irrigation on fields not registered.

Rice fields established before the year 1848 were permitted to remain; those established after that date, except as by the law provided, were subject to abolition, and their owners to fine and imprisonment.

All rice cultivated lands were to be drained in accordance with plans to be submitted to and approved by government authorities and engineers.

No new rice cultivation was to be allowed within certain prescribed limits of towns and cities of different classes, and all rice cultivated lands were to be surveyed for registration, and their healthfulness assured so far as possible by proper drainage and use of waters.

These were the chief provisions of this Piedmontese law, but there were many others which related to forms and details of administration.

The Kingdom of Italy.—It is not known by the writer hereof at this date what action the government of all Italy has taken, if any, in regard to the subject of the present section. The local regulations for Piedmont, Lombardy, and other provinces, remained in force for a number of years after the consolidation of the kingdom, and probably form at least the substance of the law on the subject to this day.

AUTHORITIES FOR CHAPTER XII.

Smith.—[Work cited as an authority for Chapter IX.] See, Vol. II, P. IV; Ch. I, Secs. II, IV, and VI; Ch. II, Secs. IV and VI, and elsewhere.

De Buffon.—[Work cited as an authority for Chapter IX.] See, Vol. II, B. VII, Ch. 38, Divs. I and IV; and Ch. 40; B. VIII, Ch. 45, Div. III; and Ch. 46.

Marsh—"Irrigation: its evils, the remedies, and the compensations." By Geo. P. Marsh (U. S. Minister to the Court of Italy, author of "The Earth, as modified by human action," etc.) See, Rept. Dept. of Agriculture, 1874.

Sardinian Code.—[Work cited as authority for Chapter X.] See, articles 599, 600, 601, 633, 551, 552, 553, 554.

Italian Code.—[Work cited as authority for Chapter X.] See, articles 575, 576, 577, 608, 536, 537, 538, 539, and remarks appended to each.

CHAPTER XIII—ITALY[5];

THE RIGHT OF WAY TO CONDUCT WATERS.

SECTION I.—*Some Ancient and Modern Laws.*
    Ancient Laws—Milan, 1216; Venetia, 1455.
    Piedmont—Code of Charles Emanuel, 1770.
    Modern Laws—Lombardy, Laws of 1804 and 1806.
    Piedmont—Sardinian Civil Code.

SECTION II.—*The Servitude of Right-of-Way for Waters.*
    Nature of the Right as a Servitude.
    Forms of the Question presented.
    The provisions of the Sardinian Code analyzed.
    The Right of Aqueduct across Lands.
    The Right of Aqueduct across other Canals.
    The Right of Aqueduct by a Common Channel.
    The Right of Aqueduct for Drainage Waters.
    References to the Italian Code.

SECTION III.—*Condemnation of Way for Works of Public Utility.*
    The provisions of the Sardinian Code.
    References to the Italian Code.

SECTION I.

THE ANCIENT AND MODERN LAWS.

SOME ANCIENT LAWS.

The necessity for a legal method for every individual to obtain in an expeditious manner the right to conduct water from a source or head of supply, across lands the property of others, and to construct and maintain works therefor on such lands, presented itself at a period very early in irrigation experience in Italy; indeed, it is probable that the realization of this point was transmitted to the Italians in some law of custom from the experience of the Romans.

"From all time the conducting of water for irrigation has been recognized as having been of special public use, which, without giving so extensive a right as appropriation, justified a notable curtailment of the rights of property."—[De Buffon, Vol. II, p. 267.

The *servitus aquæ ductus* of the Romans has reappeared in Italy as the *diritto d'acque dotto*, and, so far as known, commencing with the active extension of some great canal works, in the Milanaise province in the twelfth century, as a friendly sufferance on the part of landholders anxious to see the enterprise go on, it has developed into a well defined and thoroughly established feature of the law, in the division of servitudes established by process of law.

Although thus allowed at a very early period, the right to cross the estate of another with a canal or ditch was for a long time the subject of grave dispute in northern Italy. The several provinces were not of the same mind on the subject, nor yet were the various parts of the different provinces, united.

Milan.—Commencing, as a custom, so far as known, in the Milanaise province of the country now known as Lombardy, we find there the earliest recorded recognition of it in the form of law. This is in a code dated in 1216, which contained articles on the point, substantially as follows:

"1. Whoever has the right to obtain waters from springs or rivers, or in any other manner whatsoever, can carry it through the fields and farms of any individual, commune, or public corporation, in this state, and also across the public roads.

"2. To this end he can construct the canals or channels, and other necessary works, at the least possible inconvenience and injury to the proprietors of the farms, paying one fourth more than the true value of the land thereby occupied.

"3. In addition he must repay all damages caused by the works, according to the estimate of two practical men; provided, however, that the compensation for damages shall in no case exceed twice the value of the property damaged.

"4. He shall be bound to maintain in sufficient repair, at his own expense, the bridges and drains required for the passage of water, whether on the farms or across the roads, so that these latter shall suffer no injury, especially in rainy weather.

"5. The water may be conducted or caused to pass above or below the canals previously existing, new channels of brick and lime being made for it in such manner as that the water flowing under shall not be mixed with that flowing over, or that flowing in the preëxisting canals.

"6. The new channels must be maintained in such condition as that the proprietor of the water at the upper levels shall suffer no damage from the reflux of the same. The water shall have a free and unobstructed course."

Old as is this law, it will be seen, as this matter in hand is traced forward, that it contained all of the essential principles of a complete code on the subject, and has only been amplified, but not materially added to since.

Venetia.—In 1455 the venerable senate of the republic of Venice passed a law on this point for application in its province of Verona, whose provisions were as follows:

"Whoever shall obtain the right of establishing an irrigating channel, may demand a passage for the water across the land of any other person, paying, however, to the proprietor, twice the value of the land occupied.

"This value shall be fixed by experts chosen by the parties interested; and it shall be payable in advance, unless the proprietor of the land is willing to grant delay of payment.

"On due appraisement and offer of payment the transfer of the land is made obligatory, and should be effected by proper documents; but should the proprietor refuse, the administrative authority may adopt compulsory measures, because the right to the possession of the land for this purpose exists without reference to the inclinations of individuals, corporations, or communities, and possession obtained in the execution of the present statute shall be held good and sufficient as against the grantee.

"In the case of a proprietor refusing all acquiescence in the possession thus granted, and declining to receive the price of the land fixed as above prescribed, this price shall then be deposited with the authorities, and immediately thereafter the works of irrigation may be begun.

"When parties differ as to the proper position of the channel, the experts must always select the place least injurious to the property traversed; and the same rule must be observed in case of disputes about the location of channels sanctioned prior to the publication of this statute."

Piedmont.—Some ancient Piedmontese legislation on this subject is found in a clear form in the code of Charles Emanuel, published in 1770, as follows:

"Every commune, corporation, or individual whatever, shall be bound to grant a passage through their lands for waters legitimately derived from rivers or fountains, whether for irrigation or machinery. This passage shall likewise be granted through existing canals and water-courses, provided always that this operation shall cause no injury to the proprietors of these canals, and shall in no way impede the free course of their own proper waters.

"Whoever claims a passage for his water-course across the property of another ought to effect the same with the least possible injury. The proprietor of the water shall pay the value of the soil occupied, with one eighth in excess, as estimated by professional men. He shall further repair all damages he may cause, or pay the full value of the same.

"When a channel intersects another canal or water-course of any kind, the passage shall be effected either above or below, by means of appropriate works. The proprietor demanding passage shall be obliged to deposit security for all damages which may be caused by the said works to water-courses or canals previously in existence. This precaution being observed, the proprietor of the land cannot impede the execution of the works, but ought to lend all practical assistance during the period of their construction.

"The definite settlement of the amount of compensation for damages shall be made on completion of the works. In the event of the construction of the water-course causing a marked diminution of the extent or value of a property, the party claiming the passage shall be bound not only to pay compensation for all injuries as estimated by professional men, but also to purchase the entire property, should its owner so desire."

SOME MODERN LAWS—LOMBARDY AND PIEDMONT.

Lombardy.—Following the very ancient Milanese code, which has been transcribed under the preceding subheading as a matter of interest because of its remarkable completeness, considering the time of its production, in all times, down to the beginning of this century, the right of way to conduct water was a prominent subject for legislative and administrative consideration by the various governments and rulers who held dominion in the states of northern Italy.

Particularly was this so in Lombardy; and especially complete does the history appear to be of the various phases which the question assumed, and the steps taken in connection with it.

Of all the lines of administration, however, that which was under the guidance of Napoleon treated this subject most fully, and in the most advanced spirit. The law for the administration of the waters of the Lombardo-Venetian kingdom, promulgated by him in 1804, was the most complete and satisfactory to all parties interested that the country had ever had. And this, together with administrative decrees made under it, and dated in 1806, made up the system governing rights of way for water.

In after years (1816), when under other rule, the Austrian civil code was promulgated for this same kingdom, the good principles of the Napoleonic law, and its predecessors on this point, were overlooked, and great trouble resulted. It was considered that Lombardy had lost a most essential feature of her administrative legislation; and appeal after appeal went hence to the ruling power to reëstablish the ancient principles and regulations.

Cases wherein their absence wrought serious hardship to individuals, and detriment to the agricultural welfare of the country, were carried before the Aulic council, at Vienna; and finally, by the advice of that superior administrative body of the Austrian government, the question was set at rest; and the former law of 1804, and the several decrees on the same general subject which had closely followed it, were reëstablished by an imperial decree in 1820.

De Buffon says: "The deliberations of the Aulic council were remarkable for their equity as well as for the enlightened views expressed, including amongst other reasonings, the following considerations:

"'Running waters in this country are necessary to the nourishment of the land; they increase its fertility and assure the products. * * * Where water is so useful and contributes so powerfully to the growing of the products of the soil, doubts of its influence on the public prosperity should not be raised. * * * The new civil code should and does not oppose anything on this subject contained in the former laws and regulations. * * * Hence, under the terms of this Austrian code, they should, and do, remain in force.'"—[De Buffon, Vol. II, p. 305, quoting the decision.

The following are the provisions on this point of the law of 1804, thus re-declared to be the rule for Lombardy:

"ART. 51. Every individual is bound to cede the land necessary for the channels, the rectifications of the directions, the alteration of the courses, or the embankments of rivers, canals, or public drainage channels; and, in general terms for all works connected with water, which are designed for the public good, receiving compensation for the same at a reasonable rate.

"ART. 52. Whoever desires to make use of waters, public or private, of which he is the legitimate proprietor, for purposes of agriculture, or for the movement of machinery and hydraulic works, may carry them across the lands of others, paying the value of the soil occupied by the water-course, according to an estimate of the same, with one fourth in excess; and coming also under an obligation to maintain the water-course, banks, works, etc.; and, further, to indemnify the proprietor of the land for all damages whatsoever which the said land may sustain.

"ART. 53. Such water-courses should be carried across the portion of the estate where, according to the judgment of practical men, the least possible injury shall be caused to the proprietor, or possessor, reference being always had, however, to the convenient derivation of the water."

In addition to the foregoing, the law of 1806, also reëstablished in 1820, contained the following provision on this subject:

"ART. 16. Whoever desires to introduce water into a public canal, with the view of taking it out again at a lower point, shall submit his claim to the direction-general. It will be decided so as to cause no injury to the rights of other parties. Objections to such arrangement will be disposed of by the public administration."

These laws form the basis and principal part of all legislation on the subject in Lombardy down to the time of the promulgation of the Italian code in 1865.

Piedmont.—Following the code of Charles Emanuel III, published in 1770, and herein already transcribed, the legislation on the right of way for waters in Piedmont was contained in the Sardinian code of 1837. The very complete provisions of this code are worthy of a closer examination than those of any law which preceded it, and such examination will be given them in the next section of this chapter, where the subject is arranged for comment.

SECTION II.

THE SERVITUDE OF WAY TO CONDUCT WATERS.

NATURE OF THE RIGHT.*

The great questions which came up so early in Italy in the matter of right of way for water, were with respect to such right, exercised as a servitude: the legal occupation of one man's property by another, for the purpose of leading water across it in a canal or other conduit, without purchasing title to the property itself.

The exercise of such privilege was opposed on the ground of its being subversive of the right of property; no person, it was maintained, should have the power by simple and summary process of law, to acquire a right to continuously occupy for his purposes, any portion of the property of another. Such occupation was virtually a dispossession of one's estate in favor of another. The right of taking private property could only be exercised in the interest of the public welfare—for the purpose of public use. Conducting water for the irrigation of private estates was not a public use. The law defined what was a public use, and made provision for the condemnation of private properties, and the acquirement of title to them, when it was necessary to take them for such use. These were the arguments against the "right-of-aqueduct," as it was called.

On the other hand it was urged, that the application of water on lands so far increased their productive capacity as to make such employment a matter of great public concern and interest; that it was a general necessity in the agriculture of the country; that it could not be used without conducting it across intervening properties; that even the waters of public canals could not be distributed away from those canals without so conducting them in small private canals; that in this connection, certainly, the conducting in such small private ditches was a part of the system of the public canal, and hence a part of the necessary machinery for the public use of the waters; and that if conducting waters in a small private ditch as a distributary from a public canal was the exercise of a public use of the water, then the conducting of water from a public stream in a similar ditch was equally an exercise of a public use of it, and hence an act entitled to the privilege of occupying any property for the purpose, on making due compensation.

---
* See, De Buffon, Vol. II, Ch. 42; also, Ch. 43, p. 279, and Ch. 44, p. 307; also, Smith, Vol. II, p. 149.

FORM AND AMOUNT OF COMPENSATION.[*]

These questions were hotly discussed for centuries in Italy. As a general thing, the feudal system of land tenure was opposed to the exercise of the right to conduct water; and upon its downfall the servitude for this purpose of "aqueduct," and with it irrigation enterprise, received a great forward impulse.

It has never been proposed to take property for right-of-way for a canal without due compensation; on the contrary, the custom and law, as well, in Italy has always been to pay for the simple right of using the strip of land necessary for a canal, at its full value with the addition of a considerable percentage advance.

The facts that a canal or ditch across a property not only occupied a certain portion of its area but oftentimes occasioned its owner inconvenience, and that the presence of the water might be injurious to the land, and other similar considerations, were not lost sight of. And, furthermore, it was conceded by the advocates of the right, that the use of water for private purposes, although a necessary general use, was not a public use in the true sense.

De Buffon says: "It resulted from these considerations, that besides the recognition of the right as belonging to an irrigator, to cross with his ditch the property of his neighbor, there was stipulated in favor of the persons whose land was thus occupied an equitable regulation which aimed to make amends for the difference of taking property for public use proper, and occupying it for a purpose only indirectly for the public benefit. This rule consisted in the payment of a certain sum greater than the value of the land occupied, and the repairing of all damages occasioned by or accessory to its occupation.

"The amount of the additional indemnity, which is characteristic of the *right of aqueduct* as established by all modern nations, is variable in its nature, and has been repeatedly modified since its origin in the fifteenth century, and varied between its actual value, and twenty to twenty-five per cent advance.

"In Lombardy one pays one quarter more than the land occupied is worth; in Piedmont one fifth more, as an indemnity for damage, on values estimated in a friendly way on the opinions of experts."
—[De Buffon, Vol. II, p. 279.

Baird Smith says that, in the earliest form in northern Italy, the right of passage for waters across lands "was granted on condition that some certain supply of water should be allowed to the proprietor of the land from the canal traversing his property, in exchange for the occupancy of the soil covered by it, the use of which was temporarily lost to him."§ And he remarks that it is a curious circumstance that the same practice had been inaugurated at the earliest

---
[*] See, De Buffon, Vol. II, Chs. 42 and 43; also, Smith, Vol. II, pp. 149 and 272.
§ See, Smith, Vol. II, pp. 147-150.

stage of the modern development of irrigation under English rule in India.*

In cases where water was not allowed in exchange for the land occupied, the practice, in Italy, at first was to pay only the value of the land covered by the works, and thus rendered useless, but, as time wore on and canals became more plentiful, this bonus has ranged in some quarters as high as fifty per cent on the valuation of the lands.

FORMS OF THE RIGHT OF WAY QUESTION.

The primitive question was as to the right of conducting water across agricultural lands in a ditch; and supplementary to this came that of the right to cross with one such conduit the path of another, which acts in the early days of hydraulic works in Italy, before the art of making "syphons" and other structures to facilitate the crossing was understood, and later when such works were very expensive, oftentimes necessitated the mingling of the waters in one channel and their subsequent separation. And, then, as an outcome of this practice came up a question as to the right of one person, by paying an indemnity, to conduct waters for his benefit in the canal or ditch of another.

In addition to these three forms of the right-of-way problem, as connected with the matter of conducting water for use in irrigation, the same questions came up in connection with the right to conduct drainage waters from irrigation, drainage waters from works where such waters had been produced by other than natural causes, and drainage waters produced or accumulated naturally.

These varied natures, as to origin and purpose, of the waters to be conducted, produced modifications in the treatment which the questions have received, and in the rulings which have been made and incorporated into law on them.

These subdivisions of the subject were for the first time all fully treated in the general laws of a country by the framers of the Sardinian code, which was, in matters relating to irrigation, founded on experience in and the necessities of Piedmont.

The present code for all Italy largely followed after this model, so that I am led to present the subject upon the basis of the former law, and then, for each subdivision, point out the comparisons to be made with the latter which has taken its place.

---

* The same custom formed a feature in the outcome of irrigation development in early times in California; and there now exist perpetual water-rights, in some quarters, granted in return for a crossing of a field by a canal.

THE RIGHT OF AQUEDUCT ACROSS LANDS.

(*Sardinian Code, Articles* 622, 626, 627, 629, 640, 663, *and* 673. *Italian Code, Articles* 598, 602, 603, 605, 619, 648, *and* 666.)

The Sardinian Code.—The provisions of the Sardinian code, which specially related to the simple right of way for waters across lands, were contained in seven articles, as follows:

"ART. 622. Every commune, corporation, or individual, is bound to give a passage across their lands to water derived from rivers, springs, or any other sources, by parties having a legal right to the same, and wishing to employ it for irrigation, or for the use of works. Farm houses, with the courts, threshing floors, and gardens attached to them, are excepted from this ruling.

* * * * * * * * * * * *

"ART. 626. Whoever desires to carry water across the lands of another is bound to prove that the quantity of water whereof he is the proprietor is sufficient for the purpose to which it is destined; that with reference to the circumstances of the neighboring lands, the slopes and other conditions of the channel, the course and the free escape of the water, the line of passage demanded by him is the most convenient, and at the same time is that which will cause the least possible injury to the estates affected by it.

"ART. 627. The party desirous of carrying water over the land of another is bound to pay in advance, and before the construction of the canal is commenced, the estimated value of the ground to be occupied, without deduction of the land tax, or any other burdens which may be inherent to the soil, together with one fifth of the said value in excess, and also compensation for immediate damages, including those due to the division of the estate into two or more parts, or any other deterioration which may follow on the crossing of the land.

" In cases wherein the right of passage is claimed for any period for less than nine years, the amount to be demanded by the owner shall be limited to one half the value of the land occupied by the works, with the fifth in excess, and compensation for damages as above detailed. The claimant shall further come under obligation to restore everything to its original state on the expiration of the term agreed upon. If the party who has obtained a temporary right of passage, should desire to change it into a permanent one, the payment of the half value of the land, and the other terms annexed to the former, shall not be taken into account in settling the conditions of the latter.

* * * * * * * * * * * *

" ART. 629. In the event of the party who has obtained the right of passage for a certain quantity of water, desiring to increase the same, he shall be bound to show, first, that his canal has sufficient capacity to carry the greater volume, and, that no injury can result to the estate subject to the servitude. When the introduction of the larger volume of water requires the construction of new works, the nature and extent of these must be determined, and the value of the soil to be occupied, according to Art. 627, must be paid prior to the commencement of the said works.

* * * * * * * * * * * *

RIGHT OF WAY FOR WATERS. 255

"ART. 640. The servitude of taking water by means of a canal, or other visible and permanent work, for use in agriculture and industry, or for any other object, is included among the number of continuous and apparent servitudes.

* * * * * * * * * * * *

"ART. 663. The right of passage for water does not give to the party exercising it any right of property, either in the land at the sides, or forming the bed of the spring or water-course; and the land tax, with all other burdens attached to the soil, shall be borne by the proprietor of the aforesaid land.

* * * * * * * * * * * *

"ART. 673. The servitude is extinct if not used for thirty years."

THE RIGHT OF AQUEDUCT ACROSS LANDS—NOTEWORTHY POINTS.

The Sardinian Code.—The foregoing provisions of the Sardinian code are replete with points worthy of special notice:

Observe that the right of passage is accorded even to every individual across the lands of every other individual, municipality, or corporation; and that it is accorded for waters derived from any source whatever; but, under this law, only for the purposes of irrigation and motive power works. (See, Art. 622.)

Take notice, at the same time, however, that the right is extended only to those who have a legal right to the waters, and that, hence, in opposition to any such claim of right of way a land owner can force the would-be conducter of the water to prove his claim of right to the water itself, before he may exercise his privilege of acquiring a passage way for it. (See, Art. 622.)

Thus, the water-right claim itself was immediately put upon its merits. There could be no canal until there was a determined righ* to a definite amount of water to conduct in it; and such rights, as we have seen, could only be acquired by regular issue of privilege, or concession by the government, or by parties controlling the use of, or owning, the water; or they could result from ownership of a spring, or other water-source, such as a reservoir, or from riparian proprietorship on a stream not considered of public importance. Here, then, in these provisions on this collateral branch of the water-right question, was a safeguard against the establishment of works, and of diversions afterwards to be embroiled in litigation: the right to the water had to be proven before a right for its passage could be acquired.

Furthermore, notice the fact that the right could not be acquired for trifling and insufficient quantities of water: The "proprietor of the" (right to use the) "water" had to prove that he had a sufficiency of supply for his purpose, before he could impose upon his neighbor's estates a servitude of passage and the presence of a canal built by virtue of it.

"Very minute care was taken in the legislation of Piedmont to secure at once the efficiency of works, and the minimum of injury to lands on which they were established. Experience had shown in this country that parties frequently excavated a small well or spring on lands belonging to them; and, though the quantity of water derived from it was very trifling, they claimed the right of passage for it through irrigated fields, or in the vicinity of previously existing canals, with the view of drawing from these sources, by drainage or percolation, an additional supply at the expense of their neighbors." [Smith, Vol. II, p. 271.

The provisions which stopped this sort of imposition were contained in article 626, where we have found, also, the certain other salutary items next noticed.

The location and design of a canal or other conduit for water across the lands of another, had to be in conformity to good judgment, not only with respect to the particular service for which it was designed by its proposed constructer, but with all due regard to the continued convenient use and fruitful quality of the lands designed to be crossed; and the determination of these points was, in common with all technical and practical questions, connected with adjustments of irrigation agreements, left to hydraulic or agricultural engineers, as experts.*

"It became further clear that merely to secure the proprietor of the land from immediate pecuniary loss was not sufficient. In fixing the directions of their irrigation channels, proprietors of water might be influenced by various motives; they might desire to pass through land previously irrigated, that they might have the benefit of infiltration, or over land where there were indications of subterranean spring waters, of supplies from which their canals would derive advantage; or they might wish to benefit one neighbor by carrying water near his land, or to injure another by a contrary course.

"The government saw that it would be necessary to place limits on this freedom of choice, and hence originated the rule that prior to any special direction being determined for the canal, evidence must be laid before the competent authorities that the line selected was the least injurious to all parties concerned."—[Smith, Vol. II, p. 150.

Following out this line of policy, exacting a well determined and defined right in each case where a passage is demanded for waters, article 629 of the code, as we may have noticed, recognizes the fact that any such right was accorded only for a certain quantity of water and no more, and that when this amount was to be increased, further proceedings had to be conducted, and additional indemnity had to be paid.

This provision was made necessary by the fact that any material increase in the waters of a canal necessitate its artificial enlargement,

---

* See, Smith, Vol. II, p. 271.

endanger its banks and the adjoining lands by overflow, force its artificial enlargement by scouring out its beds and caving down its banks, or cause great additional loss by percolation into the soil of its bed and banks.

The experience of the country had made these things apparent; and it had also made apparent the fact that right of passage for any limited supply of water having been acquired and paid for, not infrequently its possessors would impose upon the land owners by forcing the flow of the canal, either for temporary convenience to supply some immediate necessity, or with the view of causing a permanent enlargement of the channel, and thereby acquire water-way for a volume of flow materially more than they had demanded and acquired the right for at first.

As will be seen elsewhere in this report, by a provision of this same code, the property owner, in consequence of this same line of imposition practiced by conductors of water, had always the right to demand that the grade plane and cross sectional dimensions of a canal through his lands should be fixed at convenient and necessary intervals along the line, by permanent and solid constructions of masonry, or other unwearing material, so that, at any time, should the canal bed be washed out at intermediate points, it could be reëstablished at exactly its former dimensions and grade by means of the guide furnished by the masonry or other solid structures along its line. These structures also served the purpose of guides by which to reëstablish the section and bottom plane of the canal when each year, in case of silt deposits having occurred, it became necessary to clear it out for the season's work.

As to conditions attached to the simple right of way for water across lands, it remains only to notice the exception to the enforcement of the servitude, which was in the case of passage across gardens, dooryards, or the sites of houses. There was a vigorous fight against this reservation in Piedmont, and, in fact, in some former laws there was no such reservation, but the framers of the Sardinian code took the view that only in cases of works declared to be of public utility—where the right of condemnation and acquirement of title to the land, should the right to dispossess a man of his house or its immediate surroundings be accorded to another.

COMPENSATION FOR RIGHT OF WAY.

The Sardinian Code.—Under the terms of this Sardinian code the proprietor who obtained a right of way for waters thereby acquired a right to the use of the strip of land necessary for the purpose, and he could devote it to no other use. He obtained no right of ownership in the land itself. The owner of the property retained this, and even had to pay the land tax on it for all time, although he had no use of it. (See, Art. 663.)

The right of the possessor of the servitude of passage was "a continuous and apparent servitude;" which meant that, unless expressly limited in an agreement, it continued for all time, even though not exercised continuously—saving the condition imposed on all such servitudes, whereby they were forfeited by non-use for thirty years. (See, Arts. 640 and 673.)

But, although the right acquired was only one of use, and not of ownership of the land occupied by the canal or ditch, the possessor of it had to pay in advance the estimated value of the land occupied, without deductions from any cause, together with one fifth of its value in excess, for the right of occupation and use; and besides that he had to pay a sum as compensation for damages to the balance of the estate crossed, by reason of inconvenience in its use, caused by the presence of the canal or ditch, or by reason of any probable injury caused to lands by seepage, or otherwise. (See, Art. 627.)

In consequence of this possible damage, also, as will be seen elsewhere, the owner of a canal had to keep it in a certain state of repair and efficiency, and to do all in his power, on demand, to prevent percolations into adjoining lands.

And not only had the would-be conducter of water to pay for his original right to cross an estate with his ditch, or canal, but he was limited to the right to conduct the amount of water stipulated and in the canal defined, and any exercise of right in excess of such prescribed privilege had to be sanctioned by a renewed negotiation and agreement, and obtained by an additional payment. (See, Art. 629.)

Finally, as may have been noticed, there were provisions for temporary as well as permanent rights of way, which were made to meet the convenience of tenants under lease of lands. Such leases were usually of nine years duration in Piedmont, and oftentimes a tenant would desire to obtain additional or other waters for a field for the period of his lease, only.

To cover these cases the second paragraph of article 627 made provision that the right might be acquired for such period or less, by the payment of one half the value of the land occupied, with the one fifth additional and the resulting damage to the lands crossed, as before explained.

It is noticeable that a temporary right-of-way could not be converted into a permanent one by the payment of the other half value of the land. Could this have been done, landlords would have taken advantage of the necessities of their tenants, to acquire permanent rights-of-way for waters for their estates, at half rates. But the law did not allow this; so that tenants were put in a position to deal on better terms with landlords, when additional supplies of water were required for an estate which they were farming.

This completes all necessary remarks on the articles of the Sardinian code relating to right-of-way for waters across lands by independent channels. It is now well that the provisions of the new code for all Italy be examined for comparison on this point, before going on to the next classification of the right-of-way matter.

The Italian Code.—The rulings of the new Italian code that take the place of those of the old Sardinian, upon which comment has just been made (articles 622, 626, 627, 629, 640, 663, and 673), are contained in its articles 598, 602, 603, 605, 619, 648, and 666, to be found in appendix II, and to which with the notes accompanying them, reference may be made in continuation of these points concerning "the right of aqueduct across lands."

THE RIGHT OF AQUEDUCT ACROSS OTHER CANALS.

(Sardinian Code, Articles 624 and 625. Italian Code, Articles 600 and 601.)

This question came up at an early period in the development of irrigation in Italy, and for a long range of time it was decided on the basis of the question of conducting water in a common channel which is next spoken of herein. The waters at a point of crossing of two canals or water-courses were taken into a common channel for a short distance, if it was desired to cross almost immediately, and then separated into two channels again by some structure designed to repartition them proportionately.

Thus this necessity for the crossing of streams or canals was probably the form which the question of a common channel took in its earliest stages, and the mingling of waters for the short space at crossings possibly suggested the mingling for purposes of economizing in conducting it for long distances.

Be this as it may, as more fully explained under the next subheading, the practice of uniting waters of separate owners and again partitioning them, gave rise to so much trouble, that the construction of special works to effect crossings of streams, without such mixing, was stimulated; and with success in this practical solution of the question came denial, in the laws, of the right to the mingling of waters; and the bare recognition of the right to cross one canal with another, in the way most suitable to the locality, and under certain restrictions, that, except in rare cases, defeat the practice of mingling altogether, is all that is left of the former right to cross as one chose.

The Sardinian Code.—The following are the provisions of the Piedmontese code on this point:

"ART. 624. It is also permitted to carry water across existing canals and water-courses in such manner as may be most expedient, and best adapted to the locality, and to the condition of the said lands and water-courses. It is necessary that the works to be constructed for the above mentioned purpose shall not stop, check, or accelerate, or in any other way change the course or the volume of the water flowing in the canals or water-courses.

"ART. 625. In carrying water across public or district roads, or across rivers or torrents, the special rules of the department of roads and waters* shall be observed."

THE RIGHT OF AQUEDUCT ACROSS OTHER WORKS—NOTEWORTHY POINTS.

The Sardinian Code.—The true meaning of the foregoing article 624 is only appreciated when we thoughtfully read the conditions attached to the apparently free right which it gives the owner of one canal to construct his channel-way across the work of another, and when we know something of the interpretations which have been given it, and the practice under it.

The passage, as we observe, must be effected "in such manner as may be most expedient and best adapted to the locality and to the condition of the canals and water-courses," but the works to be constructed in effecting the crossing "shall not stop, check, or accelerate the speed of, or in any way change the course or the volume of the water" crossed.

Now, no crossing of one flow of water by another in the same channel could be effected under these conditions: anything like a direct flow across *would* "change the course" of the current sought to be crossed; except under special arrangements involving a change of grade of such water-course, such an attempt *would* "check" its current above and "accelerate its speed" below; and any attempt at thus crossing, at all, would increase its volume and then diminish it again. Furthermore, any use of a common channel is virtually prohibited by articles 623 and 628, which are to follow under the next subheading; and the crossing of waters through other waters involves a mingling of the two, a use of a common channel, and a re-separation, even though the distance of flow together be the shortest possible.

The fact is that this article (624) was intended to do away with

---

* The civil engineering bureau of the government.

crossings involving mingling of waters, and the conditions as to the manner of the crossing and its adaption to the condition of the several works, contained in the first clause, were intended to regulate the construction of syphons under or aqueducts over these watercourses in effecting the passage.

The Italian Code.—In continuation of this subdivision of the topic attention is now asked to appendix II, where in articles 600 and 601 of the new Italian code will be found the items of the present law which supplanted those contained in 624 and 625 of the old Sardinian code upon which I have just commented, and where in the notes following these I have brought the subject of "The right of aqueduct across other canals" down to date.

THE RIGHT OF AQUEDUCT BY A COMMON CHANNEL.

(*Sardinian Code, Articles* 623 *and* 628. *Italian Code, Articles* 599 *and* 604.)

The right to make use of an existing channel or canal, in which to conduct waters—mingling them with waters of other ownership, and then reclaiming and separating out an equal or equivalent quantity, and diverting or drawing it off at some point below—is one which found place amongst the customs of the Italians at a very early period.

Lombardy.—At the time of the construction of the greatest of the ancient canal works in this valley—a truly monster canal, built during the twelfth and thirteenth centuries—there had been several other works of considerable magnitude, together with their branches, carried through the region to be traversed by the new. The art of hydraulic construction had not yet accomplished the building of large size syphon tubes under, or aqueducts for great volumes of water over existing water-courses and such works, too, would have been exceedingly expensive for the times.

Necessity, at this time, brought about the practice of uniting the waters of the old works crossed with those of the new works constructed, and then separating them again at points below, according to the gauging of outlets, or the proportioning of channels.

This is looked to as the beginning on a large scale of a practice which grew into almost a fixed custom in Lombardy, and for a long time was regarded as embodying one of the principles of the customary law of irrigation.

It never found place, as a servitude to be laid on private works, in the written statutes of Lombardy, however, and its practice brought about such disputes over the measurements on the repartitioning of waters thus mingled, that the building of structures to avoid the mingling, by dipping one canal down under the other, or carrying it around over it at points of crossing, was greatly stimulated, and the common use of channels, without common and free consent, was prohibited. Thus, article 5 of the Milanese code, transcribed in the first section of this chapter, though not explicitly so, is virtually a prohibition of the mingling of waters in a common channel.

But, in the meantime, there had grown up a very wide range of practice of this mingling and repartitioning method of crossing and conducting waters of different owners, and the annals of Italian irrigation literature are plethoric with accounts of litigations to which it gave rise. The advance in hydraulic art—in the measurement of waters—has, however, made these troubles less frequent, as the practice became more perfect and less open to objection.

The privilege of introducing waters into the government canals and then reclaiming them at a point below, was one specially open to abuse and eagerly sought after.

De Buffon, speaking of the privilege granted to individuals by the Austrian rulers of Lombardy, to introduce water into the royal canals, and then reclaim it at a lower point, says " that having all the appearance of an equitable concession, the right has hardly ever failed to degenerate into an abuse."

The absence of means and even of the possibility of accurate gauging of the amount taken in, and a just partitioning off of the amount taken out in return, and the opportunity, through the absence of continuous guarding, to take more than was due, was an incentive to the desire to obtain the privilege.

"Had such means existed at the times of which we write, demands for the introduction of private waters into the grand canals would have been less frequent.

"Justice and good sense are in accord in rejecting the idea of similarity between the simple conducting of water over the property of a neighbor, and the exercise of a right to make use of his canal already existing."

Damage caused by the construction of a canal across lands may be readily estimated and liquidated, but the injury which may be inflicted by one dishonestly inclined, upon the owner of a canal by introducing his waters therein and then reclaiming them at a point below, is a cumulative one past all finding out, and not to be estimated at all in dollars and cents alone.

"The owner of a canal, upon whom it is sought to impose such a servitude, may well reply: Our waters would be so mixed that independently of the injury you could cause me in retaking from the canal more water than you had turned in, you oblige me to keep a constant surveillance over you while doing so, and you compel me to maintain a perfect understanding with you in regard to the maintaining, clearing, and stoppage, or continuance of flow in the canal; on terms upon which we probably could not agree; in a word, you impose on me a perpetual community of interest which I have not sought, but opposed." [De Buffon, Vol. II, pp. 282–286.

The modern legislation of Lombardy contained this one provision on the subject in hand, and this was embodied in the decree or law of 1806, to which reference has heretofore been made.

"ART. 16. Whoever desires to introduce water into a public canal, with the view of extracting it again at a lower point, shall submit his claim to the direction-general. The case will be decided in accordance with Art. 4 (*i. e.*, so as to cause no injury to the rights of other parties). Objections to this arrangement shall be disposed of by the public administration."

It will be noticed that this article applied only to the introduction of water into the public canals of the State, and that it set up no basis of a right of servitude in connection with such license, but made it a mere privilege, to be extended or not extended, according to circumstances and the judgment of the superior administrative officer in charge of the works.

It formed no basis for the assertion of right to use a preëxisting private canal, and, as a matter of fact, the right has never been asserted as a servitude in modern times in Lombardy, and rarely asked, and less rarely granted, as a privilege, in the case of the government canals.

Piedmont.—In Piedmont this right first found place in the laws of the latter part of the sixteenth century, but not until the last part of the eighteenth did it assume a definite and somewhat complete form in the legislation of the country.

The edict of Charles Emanuel, published in 1770, and heretofore quoted, established the right in such broad and sweeping terms that great trouble resulted, and it became necessary to abolish it.

The Sardinian Code.—The framers of the Sardinian code, long afterwards (1837), followed out this last line of policy in the wording of the following articles:

"ART. 623. The canal required for the water shall be executed entirely at the expense of the party claiming the right of passage, and he shall have no right whatsoever to demand the said passage through canals previously in existence, and destined for the use of other waters. The proprietor of any farm, however, whereupon a canal carrying water of which he is the legal owner already exists, may prevent the opening of a new canal on the said farm, by offering to give a passage to the waters of another through the preëxisting channel, always provided that this can be done without manifest injury to the party claiming the right of passage.

* * * * *

"ART. 628. Any one availing himself of an offer made under the terms of article 623, to allow his supply of water to flow through the canal of another, is bound to pay, in proportion to the volume of water introduced by him, his share of the value of the land occupied by the works, of the excess and compensation above (in article 627) fixed, and, further, to defray in the same proportion the costs for repairs, maintenance, and every expense which the introduction of said water may have rendered necessary."

THE RIGHT OF AQUEDUCT BY A COMMON CHANNEL—NOTEWORTHY POINTS.

The Sardinian Code.—As a commentary on the last two articles of the Sardinian code, I quote the words of Baird Smith, as follows:

"The vexed question of the right of passage through previously existing channels has been very judiciously disposed of by the Sardinian legislation. To have continued this right to the possessor of water in the absolute manner established by the ancient legislation of Piedmont, would, as experience had already shown, have led to constant and harrassing disputes. The edict of Charles Emanuel, on which the right spoken of was founded, had been followed by repeated lawsuits; and though the judicial tribunals had necessarily decided all cases in accordance with its provisions, the senate of Turin had especially recorded its opinion, that the law was one of great severity.

"It is also recorded that there was scarcely ever a single case in which the result of the union in the same canal, and the subsequent division of the water belonging to two different proprietors, were satisfactory to both."—[Smith, Vol. II, p. 270.

De Buffon, also, has written of the necessity for, and justice of, this portion of the Sardinian code. Here are his sentiments:

"The power of acquiring a right of way for waters through existing canals, which, as we have seen, was admitted by the ancient legislation of Piedmont, has, for good reasons, been left out in the formation of the new code. * * * * The authors of this code found, with reason, that it was unjust to impose upon proprietors the obligations to receive strange waters into their canals, races, or ditches, as experience had proven that such mingling as resulted therefrom seldom failed to lead to litigations, disastrous to all interests."—[De Buffon, Vol. II, p. 329.

Analyzing these articles 623 and 628, we find the right of passage through the canals of another expressly abrogated—whoever acquired a right of way across lands under article 622, already considered, under this article 623 had to construct the works for carrying the water entirely at his own expense; except in the case where the owner of the land to be crossed, possessing a canal which might serve for the accommodation of the waters desired to be conducted, might, in order to save his land from being cut by another channel, offer to accord the right to use his canal.

Then article 628 makes provision for his compensation, if his offer is accepted, for the use of his work and the occupation of his land.

The Italian Code.—Following this subject of "The right of aqueduct by a common channel," I now invite attention to articles 599 and 604 of the new Italian code, found in appendix II, and which took the place for all Italy of the foregoing 623 and 628 of the old Sardinian code for Piedmont. And also, in notes to the Italian code articles will be found a continuation of the remarks already made on the points of this subheading.

THE RIGHT OF AQUEDUCT FOR WATERS OF DRAINAGE AND FOR WARPING.*

(*Sardinian Code, Article* 630. *Italian Code, Articles* 609, 610, 612.)

The right of passage for waters of natural drainage has always been recognized as a servitude over lower lands, due to their situation; but the right to discharge natural drainage waters on to lower lands at points other than where their original course led them, or to increase the volume of drainage waters artificially, has been the subject of legislation.

Lombardy.—The chief part of the modern laws of Lombardy on this point were embraced in article 54 of the decree of 1806, which first declared that "lower lands are bound to give passage to waters flowing from higher levels." It will be noticed that this was a declaration of a right of servitude, not only in favor of natural drainage waters, as was the case in other codes spoken of, but it included all waters draining from higher lands, thus including waters which may have been caused to flow down by artificial means, or which had been brought to the lands by artificial works.

Not, however, to injure the owners of such estates as might be subjected to this servitude, the second paragraph of this article contains the following provision: Referring to articles 51, 52, and 53, which relate primarily to rights-of-way for waters for irrigation, but are also made to apply to rights for drainage waters, and which contain obligations on the proposed conducter of water to pay for the privilege, etc. This article 54 further says:

"In addition to the obligations imposed by the preceding articles, the proprietor of the upper lands is bound to defray the cost of such drainage channels as may be necessary, and of such works of defense

---

* See, De Buffon, Vol. II, pp. 194–200, and 333; also, Smith, Vol. II, pp. 158, 268, and elsewhere.

as may be required to protect the lands through which the waters pass; as, also, to repair any damage which at any time the lands may sustain. The preceding article does not affect special agreements between proprietors, nor rights of servitude acquired by process of law."—[See, Smith, Vol. II, p. 158.

This above is an example of most extended application of the servitude of drainage from higher to lower property. Seeing that it includes the right for all waters from the higher estate, whether naturally running off or not, in with those produced naturally, and puts the burden of receiving them upon the lower estate, as a natural servitude.

In Italian the word *colatura* is used as a name for surplus waters which have been used in irrigation—the drainage waters from irrigated fields; and where rice and meadow lands are cultivated the quantity of these *colaturas* is very large, amounting in some cases to half as much water as is originally applied to the land.

The right of passage was accorded in Lombardy for *colaturas* upon the same terms as the original waters for irrigation. They were themselves used again in irrigation, and were held to differ in the eyes of the law in no respect from waters directly derived from a primary source, such as a river or spring.

Piedmont Sardinian Code.—The following article of the Sardinian code embraced the modern legislation of Piedmont specially applicable to the matter of "The right of aqueduct for waters of drainage and for warping:"

"ART. 630. The terms established in the foregoing articles for the passage of water apply equally to the case of the proprietor of a marshy estate, who desires to improve the same either by the process of warping (*colmata*) or by the excavation of one or more channels of drainage.

"Should opposition be made to the estate by parties having rights to the water on or flowing in any way from the same, the tribunals, in deciding, ought to have due regard to sanitary and agricultural interests, and also to the use made of the water by the objecting parties."

Herein we notice the importance attached not only to drainage, but also to the process of *colmatage*, spoken of on page 82, *ante*, and elsewhere, that they should be placed on the same footing with irrigation in the matter of acquiring right to passage for waters to effect their purpose.

" It is held that the drainage or the improvement, by the process of *warping**, of marshy localities has an influence on the general good of the community scarcely inferior to that of irrigation itself, and that he who is prepared to invest his capital in changing miasmatous

---

* *Colmatage* (French), or *colmata* (Italian).

swamps into fertile fields is entitled to privileges at least equal to those afforded by the laws to the proprietor of water employed in increasing directly the products of agriculture and industry."—[Smith, Vol. II, p. 268.

The second paragraph of the foregoing article (630) applies to cases where rights to the use of drainage waters drawn from lands-in-a marshy state, which have accrued by use or otherwise to the benefit of some one situated below, are to be interfered with by the owner so improving the estate by more thorough drainage as to intercept, stop, or change the outfall of the drainage waters; and the existence of the law is an evidence of the thoroughness of the system which thus usefully employs in irrigation even the waters of marshes and swampy tracts to an extent that would require and bring about the passage of a special clause in the protection of such rights.

The right of way for drainage waters from irrigation (*colaturas*) is also assured by the provisions of this Sardinial code, seeing that the privilege is accorded for all waters to which a legal right is had by the person who desires to conduct them, and seeing that these waters are devoted to re-application in irrigation or some other useful purpose.

The Italian Code.—In conclusion of this subject of "The right of aqueduct for waters of drainage and for warping," the attention is now directed to articles 606, 609, 610, 611, 612, 637, 654, 655, and 656,* which contain the relevant matter of the present law, and also to notes relating thereto, which are in conclusion of the remarks already made under this subheading.

SECTION III.

RIGHT OF AQUEDUCT FOR PUBLIC WATERS.

CONDEMNATION FOR PURPOSES OF PUBLIC UTILITY.§

(*Sardinian Code, Articles* 441 *and* 442. *Italian Code, Article* 438.)

Articles 622 to 629 of the Sardinian civil code, as we have seen, provided for the establishment of a right of way for waters through the estates of others. The right to be acquired under these articles was a simple servitude upon the property crossed, and did not give the owner of the canal or water any right of ownership in the strip of land occupied by his work. The application of these principles was intended to provide for cases of the most insignificant kind, even where any one had a piece of land large enough to make a garden patch worth irrigating, and water enough to take to it for the purpose.

---

* See, Appendix II.
§ See, De Buffon, Vol. II. pp. 333–337; also, Smith, Vol. II, pp. 274–278.

In the construction of great works, whether of a private or public ownership, intended to irrigate considerable areas of land, the Sardinian government recognized the presence of the element of public utility in the project, and provided in its code for the acquirement by condemnation, or the exercise of what we term the power of eminent domain, of an absolute property right in the lands occupied by the canal and its necessary accessory structures. The following are the provisions spoken of:

"ART. 441. No one can be compelled to cede his property, or to allow another to make use of it, unless for objects of public utility, and on receipt of a just compensation, payable in advance.

"The works of public utility, and the property to be occupied by the same, are determined under provisions emanating from the sovereign.

"The rules to be followed in the aforesaid cases shall be fixed by special laws and regulations.

"ART. 442. When the parties cannot agree before the administrative authority on the account of the indemnity to be paid, the disputes shall be decided by the legal tribunals."

Following this, and in 1839, a long special law was passed regulating this matter of expropriation of private property for purposes of public utility, the chief provisions of which are as follows:

"ART. 1. Works of public utility are those executed on account of the royal domain, of the state administrations, of provinces, and of communes. Such works, and the property to be taken possession of in the execution of the same, shall be determined under article 441 of the civil code, by letters patent issued under the advice of the council of state.

"ART. 2. Works executed by associations or single individuals may also be declared of public utility by appropriate letters patent, whenever their importance or their influence on the development of the general prosperity is such as to warrant this character being attributed to them."—[See, Smith, Vol. II, p. 274.

This law made provision in minute detail for the conducting of the sale of the lands to be occupied by the works: First, prescribing in thirty-seven articles the steps and conditions of procedure when the arrangement could be amicably consummated between the parties, with the mediation of the intendants of the provinces; and then, in twenty-three additional articles, specifying the forms of procedure, etc., to be had before the courts in carrying out the enforced sale, if an amicable agreement could not be arrived at.

The general object is to insure a fair compensation for the ground occupied, and for the injuries or inconveniences occasioned by its use for the purpose desired, to be estimated upon the basis of the returns from the land for the ten years preceding the date of the proceedings, or on other equally exact data for the purpose.

The first estimates were to be made under the direction of the administrative authorities; but upon appeal to the courts, if demanded by either party at interest, new estimates were made, under the direction of the courts, by engineers, selected amicably by the disputants, if possible, or, otherwise, appointed by the court itself.

This law was commented upon and put in operation by a number of administrative regulations, which further explained the forms of procedure to be had and the application of the articles of the law. As to just what construction was to be placed on the term "public utility," the following clauses are in point:

"Whenever the limit of simple private interest, whether in the case of corporate or individual proprietors, is passed, and the work is designed for public service, the declaration of public utility may be claimed without reference to the special nature of the work itself.

"Therefore, canals and ditches, provided they have a material influence upon public prosperity, become included among those works, on behalf of which the declaration of public utility may be made.

"In fact, although the civil code has established certain special rules for canals and ditches, with the view of facilitating their construction by private parties, it does not thence follow, that, when they present all the characteristics of other works of public utility, the benefits secured by the laws to such works should be denied to them.

"It is, therefore, to be concluded that when projected canals have the characteristics of works of public utility, they shall be so declared, with the view of applying to them the provisions of the law on dispossession."—[Trans., Smith, Vol. II, p. 277.

These articles of the laws and administrative regulations so very fully and plainly explain this branch of the subject that comment is unnecessary.

To review, in one paragraph, the bearing of all that has been said about the right-of-way laws in Piedmont, it may be said (1), that any person, association, or corporation, having a valid right to any certain amount of water sufficient for and to be applied to a declared useful purpose in agriculture or in the creation of power by water-power works, could acquire the use of the strip of land necessary upon which to construct a canal or other conduit to convey the water from the place of its source to the point of its intended use, as a simple servitude, by observing due forms of demand, paying just compensation for lands and damages in advance, and engaging to so use and maintain his works as to cause no avoidable damage to the owners of the lands traversed; and (2), that when the use of the water was such as to be of public benefit, authority might be had wholly to condemn, pay for, and acquire title to the strip of land necessary upon which to locate the work; and (3), that "public utility" in the case of irrigation meant the use of waters by a number of individuals, a neighborhood or community of irrigators.

Of this Piedmontese system De Buffon wrote:

"It takes but little reflection to realize that the modern legislation of Piedmont on the subject in question, is as complete and satisfactory as could be desired. Furthermore, the results of it are more convincing than reflection on the laws themselves could be. It may be said that agriculture cannot hope to be more truly and efficiently protected than it is in this country, especially by the facilities which legislation has given for the extending of irrigation."—[De Buffon, Vol. II, p. 337.

The Italian Code.—In conclusion of this subject of "Condemnation of right of aqueduct for works of public utility," attention is asked to article 438 of the Italian code, in appendix II, and to the remarks following it.

AUTHORITIES FOR CHAPTER XIII.

De Buffon.—[Work cited as an authority for Chaper IX.] See, Vol. II, Book VIII, Chaps. 42 to 45; Book IX, Chap. 47, Div. II, and Chap. 48; also, See Vol. I, Book II, Chap. VII, Div. I.

Smith.—[Work cited as an authority for Chapter IX.] See, Vol. II, Part IV, Chap. I, Sec. 2; Chap. II, Sec. 2, and elsewhere in Vols. I and II.

Sardinian Code.—[Work cited as an authority for Chapter X.] See, Arts. 441, 442, 622 to 630, 640, 663, and 673.

Italian Code.—[Work cited as an authority for Chapter X.] See, Arts. 438, 598 to 605, 609 to 612, 619, 648, and 666, and remarks appended to each.

CHAPTER XIV.—ITALY[6];

IRRIGATION ORGANIZATION AND REGULATION.

SECTION   I.—*Irrigation Organization.*
　　　　　Causes of and Necessity for Organization.
　　　　　Social Tendency of Irrigation in Italy.
　　　　　Formation of Irrigation Associations.
　　　　　General Law in Lombardy.

SECTION II.—*Organization and Management.*
　　　　　The General Association West of the Sesia—Piedmont.
　　　　　General Organization and Management.
　　　　　The Government and the Association.

SECTION III.—*Organization of Associations.*
　　　　　The present Law for all Italy.
　　　　　Voluntary Association of Landholders.
　　　　　Compulsory Formation of Associations.

SECTION I.

IRRIGATION ORGANIZATION.

CAUSES OF AND NECESSITY FOR ORGANIZATION.*

As heretofore written, the canal works of Italy, for the most part, had their growth in times when only rulers, governments, rich civil and ecclesiastical corporations, wealthy nobles, and large landholders could afford to undertake such enterprises. There was no such thing as companies or associations of small farmers taking water out on their own account for their own use. The river channels and topography and climate of the country were such that the diversion of waters necessitated the construction of great works, built solidly of costly materials, at enormous expense; so there were but few small private canals, and no cheap works of diversion from the natural streams, such as it has been possible to construct and maintain in California.

Then again, the landholdings were much consolidated into few hands in years long gone by, and the practice of irrigation has

---
* See, De Buffon; also, Smith, generally.

tended to make the rich richer and the poor poorer—to further increase the size and diminish the number of farms, and reduce the farm workers from the grade of small landholders to those of renters of and laborers on other people's property.

Thus, the canals are owned by the government, by wealthy nobles, ecclesiastical and municipal corporations, and very generally not by the irrigators, and the waters are commonly sold by volume according to established units of measure and standards of measuring devices.

The lands are generally held by rich non-working owners, and are farmed by irrigation under the management of professional superintendents, or are divided into small tracts and leased, for terms, generally, of nine years duration, to the real working irrigators of the country.

SOCIAL TENDENCY OF IRRIGATION IN ITALY.

Besides the authors, after whom, in a general way, I have written the foregoing paragraphs, the writings of another deserve special consideration on this point.

I quote from a paper written by the late Hon. George P. Marsh, for many years United States minister at the court of Italy, an author of learning, observation, and thought in the special line of the physical, social, and moral effects of man's greater occupations on the earth's surface.

Speaking of the effects of irrigation generally in Europe, but from personal observation, more particularly, in Italy, where the paper was written, this author says:

"With an important exception,* which I shall notice hereafter, the tendency of irrigation as a regular agricultural method, is to promote the accumulation of large tracts of land in the hands of single proprietors, and consequently to dispossess the smaller landholders."

He then gives some reasons for this, which I shall desire to present in a later part of this report, so will not transcribe them here, but I go on with his narration of the facts to our point :

"European experience shows, as might be expected from what has just been said, that under such circumstances, as well as where waters belonging to the state are farmed and relet by private individuals, water-rights are a constant source of gross injustice and endless litigation. The consequence of these interminable vexations is that the poorer, or more peaceably disposed landholder, is obliged to sell his possessions to a richer or more litigious proprietor, and the whole district gradually passes into the hands of a single holder, or family, or corporation. Hence, in the large irrigated plain lands of Europe,

---

*This exception is in the case of several of the irrigation regions in the south of Spain, where the water rights have been held by the peasant proprietors from the time of the Moors.

real estate is accumulated in vast tracts of single ownership, and farming is conducted on a scale hardly surpassed in England, or even on the boundless meadows and pastures of our own west. * * *

"The small cultivators who sell their paternal acres must either emigrate, and so diminish the resident population, or sink from the class of land owners to that of hired laborers on the fields which, once their own, are their homes no longer. Having no proprietary interest in the soil they till, no mastership over it, they are, as I have said elsewhere, virtually expatriated, and the middle class, which ought to constitute the true moral as well as physical power of the land, ceases to exist and enjoy a social status as a rural order, and is found only among the trading and industrial population of the cities."—[Marsh; see, Report Department of Agriculture, 1874, pp. 364–366.

Although this picture of experience is in a general way true for the irrigation regions of central Spain, of France, of Belgium, and other central European countries where irrigation is practiced on a large scale at some notable localities, it is specially true for Italy and particularly of Tuscany and the regions of the valley of the Po, which we are now considering.

With these antecedents and this tendency of canal enterprise and irrigation practice, grew up the gravest conflicts between the irrigators and the holders of water-rights—the canal owners. The small land proprietors and irrigators found that singly they could not hold their own: they must organize locally, and, as a body in each neighborhood, district, or subdistrict, treat with those who supplied them with water. This necessity brought about the demand for a law prescribing a form for such organizing, and recognization as legally constituted bodies. And, hence, as a modern outcome, we find the law which I am now about to transcribe.

FORMATION OF IRRIGATION ASSOCIATIONS—LOMBARDY.

The association principle, of which this codification was the outcome in Lombardy in the early part of the present century, is of quite ancient origin, but its full development in the form we now find it was due to the organizing genius of Napoleon, under whose dictation the original text of the Lombardian law was promulgated. It was a general law governing the organization of land owners, for purposes of drainage, reclamation, or irrigation. And its application, as will be seen, was committed largely to the government administrative authorities.

I.—Organization of Associations.

"1. All proprietors interested in special hydraulic works shall be formed into such number of associations as may best suit their common interests and the territorial divisions of the kingdom.

"2. All existing associations shall be preserved, with such modifications or additions as may appear desirable.

"3. The list of associations shall be definitely published during the course of the following year.

"4. The associations are subjected to the control of the prefecture, and shall exercise their functions according to such rules and regulations as may be prescribed by the superior authorities.

"5. The properties benefiting by one drainage or irrigating canal constitute a district.

"6. All the proprietors of estates situated within a district, constitute an association.

"7. If the extent and circumstances of a canal should so require, it may be divided into several sections; each section may have its own district, and each district its own association.

"8. Each association shall be represented by a delegation.

"9. The number of delegates shall be determined by the direction-general, in proportion to the wants of the district.

"10. The proprietors in each district shall elect the members of the delegation by ballot. To this end the prefecture shall convoke the proprietors at a specified time and place. The assembly shall be presided over by the prefect, the vice-prefect, or one of their deputies. If the number present shall not equal a third of the total number of the proprietors, those actually present shall select the delegates from three lists composed of the larger proprietors.

"11. One delegate shall be removed from the delegation biennially. The retiring delegate shall be selected by lot from among those first elected; afterwards the senior member shall retire. The retiring delegate may be reëlected indefinitely.

"12. The delegation has a president, whose tenure of office lasts for one year. All the delegates succeed to the presidency in due order. Among those first elected, the majority of votes in the election shall regulate the order of succession. Subsequently, the rule of seniority shall be observed.

"13. The delegation shall determine the days of its ordinary meetings. The prefect, the vice-prefect, and the president can, on necessary occasions, summon extraordinary meetings. The president shall cause the decisions of the delegation to be executed in all cases where no special member has been nominated for this purpose.

"14. The ordinary duties of the delegation are to superintend the canals with their outlets and banks, as also the works of such other canals as may traverse or surround their district; to maintain all these in repair, and to collect the funds necessary for these objects.

"15. The delegation shall decide on all matters within its powers by simple majority of votes.

"16. When new projects interesting to the entire district come under discussion—such as the construction of new canals, the enlarging or prolonging of old ones, the formation of outlets or tunnels under rivers, or similar works involving extraordinary outlay—then the whole of the proprietors of the district shall be convoked, and shall proceed to the election of as many extraordinary as there are ordinary delegates.

"17. The union of the additional with the regular delegates forms an extraordinary delegation, which shall decide on the proposed works, and the means of executing them.

"18. The result of the deliberations of the extraordinary delegation shall be submitted for approval to the direction-general. On the works and means of execution receiving the approval of the superior authorities, their completion is intrusted to the ordinary delegation.

"19. Each delegation shall have an accountant and a cashier.

"20. In such districts as have relations with foreign powers, the conventions and customs in present force shall continue.

"21. In cases of new canals, or improvements of land by drainage or warping, the districts and associations shall be formed in accordance with the foregoing rules.

II.—Superintendence of Works.

"23. There shall be nominated for the superintendence of the canals, outlets, and embankments, belonging to a district, such number of guards as the delegation may consider necessary.

"24. The delegation shall prescribe police rules for the regular protection of these works.

"25. The ordinary engineer shall visit triennially, and oftener if requisite, all the canals in his department. He shall examine the interior condition of all the fixed works; note their wants, defects, or abuses; propose to the delegation the appropriate repairs; and shall inform the engineer-in-chief of the whole, who will then report to the direction-general. Should the delegation not be prepared to execute the works suggested, the engineer-in-ordinary will report accordingly to the engineer-in-chief, who will then submit the question to the direction, with his observations and opinion upon it, for the decision of the superior authorities. During such visits the condition of new land improvements should be especially noted.

"26. In times of floods, or inroads of waters, the extraordinary guard reserved for such occasions shall be bound, in operating on any works belonging to any particular district, to act in accordance with the wants and usual customs of the localities.

III.—Works on Canals of Drainage.

"27. With the view of showing clearly the interior condition of the principal canals, fixed marks shall be established at every fourteen hundred feet along their banks, on which shall be shown the depth that each section of the channels ought to have. This depth shall be shown in local measures, with the equivalent Italian measures, on each of the marks above referred to.

"28. Each delegation shall fix a certain minimum depth for each canal within its district; and when silting up above this plane takes place, recourse should immediately be had to excavation. The depth in question should be approved of by the engineer-in-chief.

"29. The clearances of the canals should be effected twice a year, at least.

"30. Should it happen that, by any river breaking its embankments, portions of canals are blocked up, the delegation should instantly reëstablish the same.

IV.—Distribution and Collection of Funds.

"31. A preparatory estimate of the expense required for the public and communal canals included within a district shall be made by an engineer or other qualified party. The same course shall be followed in all cases of extraordinary works.

"32. The delegation shall prepare annually an assessment list, the basis of which shall be the amount of annual public burdens on each property, and the estimates of the probable expense required for the works, as given by the engineer.

"33. This assessment list shall be submitted for the approval of the prefect, who shall forward the same for the consideration of the magistracy of waters. On receiving the sanction of the preceding parties, the assessment shall be enforced according to existing agreements and customs.

"34. Where no special agreements or common customs exist, the proprietors subject to the assessment shall be arranged in different classes, according to the amount of benefit they derive respectively from the works.

"An engineer-in-chief, selected by the president of the delegation, shall propose the classification of the proprietors and the different proportions in which the separate classes shall contribute to the expenses.

"This proposal shall be made public, so that the proprietors may present any objections that they may have to it, before the provincial delegation within a term to be fixed by this body. The provincial delegation, with the concurrence of the provincial assembly shall report on the case to the government. On receipt of the approval or alterations of the superior authorities, the quota fixed for each class of proprietors shall be distributed among the individuals composing this class, in proportion to the revenue survey valuations of their respective properties.

"35. In collecting this assessment the cashier shall exercise the same powers as are prescribed by the laws for the collection of the direct taxes.

"36. The fines imposed on parties breaking the existing regulations belong to the association, and shall be deposited in the treasury of the same. Whatever profits may in any way accrue shall be similarly deposited in the treasury, at the disposal of the delegation.

"37. The cashier shall make payments on orders signed by the president, one delegate, and the accountant. He ought to be required to furnish a sufficient amount of security. He is appointed by the presiding body, on its own responsibility. The entire amount of each rate imposed shall be placed to his debit five days after it has become due, whether it has then been received by him or not.

"38. At the end of each year the superintending body shall present to the provincial delegation the accounts of the expenditures, with a statement of the debits and credits of the treasury; and when these have been approved by a vote of the provincial assembly, they shall be published, and a copy forwarded to the government.

"39. In case of several channels, which cannot conveniently be included within one district, having a common escape-canal or other works, the expense required for the protection and maintenance of the said works shall be divided among the districts using them, in proportion to their respective interests in them, excepting always any agreements in force to a different effect.

"40. If the defense of an embankment concerns several districts, the expense of repairing it shall be divided among them according to their respective interests, saving agreements in force to the contrary.

V.—General Provisions.

"41. Associations of proprietors interested in drainage, land improvement, or irrigation, are subject to the inspection of the provincial delegations, and are placed under the guardianship of the political administrative authority. They exercise their duties according to the rules and regulations prescribed by the superior authority.

"42. All works appertaining to associations shall be made by regular contracts. To proceed otherwise, and to execute the works by daily labor, requires an express order from the government, who will decide on the case and the necessity. In contracting for the annual repairs of the works the contracts shall be made for nine years. The government may alter this arrangement under special circumstances.

"43. The channels shall be furnished, not only with the appliances necessary for opening or closing them with facility, but also with supplies of all the materials which may be required to strengthen and protect them in time of floods. All arrangements that concern the defense of embanked rivers are under charge of the engineer-in-chief and his subalterns.

"44. Where the respective titles do not otherwise provide, the volume and the special regulations for each outlet from the rivers shall be fixed in such manner as that no injury may result to the interests of any of the proprietors belonging to the district. The same care shall be observed in the use of turbid waters employed in operations of improvement by deposits.

"45. Objections made by the proprietors within a district to the proceedings of the presiding body shall be considered by the provincial delegation which, having verified the facts and submitted them to the provincial congregation, shall decide each case according to its merits. If the objections should involve points of great importance the provincial delegation shall submit them to the government, and shall await its instructions before coming to any decision.

"46. Each delegation shall present to the provincial delegation a project of regulations for the careful protection of all the matters committed to its charge. These regulations shall not have force until approved of by the protecting authority."—[Smith, Vol. II, pp. 170–178.

SECTION II.

ORGANIZATION AND MANAGEMENT OF IRRIGATION ASSOCIATIONS.

THE GENERAL ASSOCIATION OF IRRIGATION WEST OF THE SESIA—PIEDMONT.

To illustrate the application of the formation of irrigation societies in Piedmont, I bring forward the special case of the organization and operation of the General Association of Irrigation west of the Sesia.

As already written, all of the canals, of which there are quite a large number, in a certain great district east of the Po and between the Dora Baltea and Sesia 'rivers in Piedmont, were the property of the government. These were managed and maintained, as explained in the preceding chapter, under the direction of a bureau of civil engineering, attached to the ministry of finance; but the waters were farmed out on leases to contractors who undertook to distribute them to the irrigators, collect the rents, and pay the government specified sums annually.

The arrangement is thus spoken of:

"In the hands of the farmer of the canal revenues, are vested the powers of entering into the contracts for water with all cultivators—of fixing in communication with them the annual rent to be paid, and the manner in which the supply of water is to be used and measured. In a word, the whole interior economy, so far as the granting of water is concerned, is under the control of this party, who has his own private agents spread over the country to watch his interests and carry into execution his orders, all disputes being submitted to the decision of the ordinary tribunals."—[Smith, Vol. I, pp. 119–122.

In view of this state of things, and remembering the matter upon this special point—the effect of irrigation on land holdings—transcribed from the writings of Marsh, in the first section of this chapter, we see the incentive to the organization of the irrigators of this region, and are prepared to appreciate the importance of their association, which was formed for the purpose of doing away with the evils of the system of water-leasing by government to contracting "middle-men," and re-leasing or selling to the irrigators.

The general idea was the formation of a society composed of all the irrigators, themselves to lease the waters in bulk from the government canals, and distribute them to themselves as irrigators.

The society was founded by government under the act of July 3, 1853, and owes its origin to count Cavour, at the time minister of finance of the Sardinian government, and a man to whom, on account of his liberal and advanced ideas and patriotic actions, northern Italy looks as a benefactor.

The organization had for its object at starting, "to lease, administer, and employ in general, according to an economical and natural system of irrigated cultivation, the waters of the crown canals derived from the Dora Baltea (river), in terms of the agreement made with the state finance bureau, for the irrigation of the respective properties of the shareholders, with the power of extending successively the benefits of the association even to the mutual assurance against losses by hail, fire, and such like, and to other social objects of mutual profit."

By the agreement referred to, the government leased to the association all of the waters, in volume about 1,750 cubic feet per second at maximum flow, of the crown canals in the region spoken of, which is about twenty miles square, for a period of thirty years; making, however, certain reservations in favor of owners of old rights long ago acquired in perpetuity, which reservations amounted to an aggregate volume of 793 cubic feet per second. The lease took effect on January 1, 1854, expired the first of the present year, and has been renewed on substantially the same terms.

I have made and hereinafter present an analysis of this agreement, but now ask attention to an account of the organization and internal working of the society.

The statutes and regulations of the society comprise three hundred and seventy-nine articles, covering seventy-six pages octavo. The following is an abstract of the principal points contained in these:

ORGANIZATION AND MANAGEMENT OF THE ASSOCIATION.

"In each *commune*, or parish, irrigated by these canals, there is a society termed a *consorzio agrario*, composed of all the proprietors within the parish who take water for their lands; or in certain cases a *consorzio* may be composed of proprietors of adjoining small parishes. Each *consorzio* elects by universal suffrage one or two deputies, according as it uses a discharge of less or more than 30 modules (61.4 cubic feet per second) on its irrigation. These deputies form an assembly for the general administration of affairs. They must be themselves members of the society, over twenty-five years of age, "sufficiently acquainted with agriculture," and men of good character. They receive no salary as deputies, nor are they allowed to hold any paid office under the society. They are elected for three years, and may be reëlected. They meet regularly twice a year, on the fifteenth of March and fifteenth of November, and half their number form a quorum. They elect from among themselves a president and vice-president, whose functions last for three years, and each year they choose also an honorary secretary and two assistants. They pass the accounts of the year, settle how much is to be paid by each *consorzio*, what salaries their employés are to have, listen to suggestions for the benefit of the society, and in short, generally direct and control the whole of its business. The rules passed by the assembly are binding on all the members of the society. To help them in forming decisions they have a legal and an engineering adviser.

THE DIRECTION-GENERAL.

"From among themselves the assembly elect three committees: the direction-general, the committee of surveillance, and the council of arbitration.

"The first is the committee of management of the affairs of the society. It consists of a director-general, three members, a secretary, and an assistant secretary. If the director-general likes he may appoint a colleague, with the approval of the assembly, to take his place in case of illness or absence. The director-general may call on the assembly to dismiss any of the members of his committee, or he himself may suspend them for not doing their duty. He has in every way to watch over the interests of the society, to see to the conduct of its servants, and to give them rules for their guidance, to direct any works, to disburse expenses, to arrange with the government (or with the canal company) for the amount of water required at each point, to see generally to the distribution of the water over the irrigated district, to carry on all communications with the government, and in short to be general manager. The director-general receives an allowance of $1,800 a year, from which he is expected to pay a number of small charges; and each member of his committee receives a certain salary. This committee has its headquarters at Vercelli, and renders an account of its proceedings at each meeting of the general assembly.

"The committee of surveillance is 'the eye of the assembly over the direction-general,' and has to see that it carries out faithfully its duties towards the society. It consists of three members, of whom the oldest presides. They meet once a week, and each time receive a ticket which entitles them to a small allowance as fixed at each general assembly; in 1866 the whole amount being only $170. Should they think necessary, they may call an extraordinary meeting, and make a report of their proceedings.

THE COUNCIL OF ARBITRATION.

"The council of arbitration has for its object: 'first, to settle all disputes regarding affairs of the society which may arise between the members and the society, or between the society and its servants; second, to decide cases of breaches of the rules and discipline of the society; third, to assist the society in actions before the courts; fourth, to give their advice on whatever may be referred to them by the director-general; fifth, to fix and settle in case of dispute the compensation for the passage, outlet, or any other obligation or damage occasioned by the flowing, distribution, employment, recovery in drains, and escape of the waters of the society, with its members or among the *consorzios*, or members with each other.' This council is composed of three members of the assembly, who must be resident in Vercelli, and are elected annually. They receive no regular pay, but get certificates of attendance at meetings, like the committee of surveillance, and these certificates entitle them to a small remuneration, of which the whole amount in 1866 was $243."

Their decisions are settled by the opinion of the majority. There is always the power of appeal from them to the ordinary courts of justice; and to admit of this appeal, the execution of their sentences is deferred for fifteen days after being promulgated, unless in cases where, for the sake of the crops, it must be carried out at once. After fifteen days, if no appeal has been made, the decisions of the council are looked on as final. When necessary the council summon a lawyer or an engineer to their assistance. All charges of this council are paid by whoever loses the case. The director-general is not allowed to carry on any lawsuit on the part of the society without the previous sanction of the council of arbitration.

FINANCE AND SUPERINTENDENCE.

"The money transactions of the society are under a cashier, who has to give a security for $4,000, and who is responsible for all connected with the cash. His chest has three keys, of which he keeps one, the director-general another, and the third is held by the largest shareholder of the society who is a member of the general assembly and happens to live in Vercelli. Money is issued on the checks of the director-general, and once a month he and the member who keeps the third key of the cash-box count the cash, and audit the cashier's books.

"To effect the distribution of the water, the area irrigated is divided into a certain number of districts (at first only four but increased since), in each of which there is an overseer in charge of the irrigation, termed the *delegato*, who receives his orders from the director-general, and several guards or water-bailiffs, termed *acquaiuoli*. These officers patrol the water-courses; see that the modules are discharging their proper amount; that the water that passes off the fields is not running to waste, but is caught in the catch-water drains, from which at a lower level it can be again utilized (a point attended to with admirable care in the Piedmontese irrigation), and do all the other ordinary duties connected with their position. Neglect of duty or disobedience of orders subjects them to fines, reduction of salary, or dismissal."—[Moncrieff, pp. 230–234.

THE GOVERNMENT AND THE ASSOCIATION.

For the purpose of bringing out as clearly as possible the relations existing between the government and this great irrigation association, I have made an analysis of the agreement between them, and, doing away with superfluous verbiage, have brought the essential points together under the headings which have seemed adapted to the matter and calculated to convey the best idea of its scope and bearing, as follows:

An Analysis of the Lease of Waters to the "General Association of Irrigation West of the Sesia," Piedmont, 1854.

WATER RIGHTS AND PRIVILEGES.

The government granted to the association, for a period of thirty years, the exclusive right to the use and control of the waters of the three large state canals, Ivrea, Cigliano, and Rotto, derived from the Dora Baltea river, to the extent of such volume as might be necessary to properly irrigate the districts of Vercelli, Casale, and Biella, wherein were situated the lands owned by the members of the association.

At all times the volume of water to be delivered was to be limited by the total capacity of the canals, which was 870 modules, or 1,750 cubic feet per second, and by the previous engagements of the State to deliver, of this amount, 387 modules to holders of old grants-in-perpetuity of water-rights.

The supply of water furnished was to be negotiated for each irrigation season in advance—the summer season being held to commence with the spring equinox and end with the autumn equinox, and the winter season to embrace the balance of the year—and, with the exceptions noted, the supply during each such irrigation season was to be delivered by the three canals in proportion to their respective capacities, and maintained at a steady flow equivalent to the amount engaged.

During the summer—the season of abundant supply and the season of greatest demand—the volume delivered was to be governed by the demand of the association made before the thirty-first of January preceding, and by the limitations already spoken of.

The government reserved the right of closing the canals for necessary repairs according to custom during the spring and before the first of April, at which date, unless some extraordinary obstacle or reason prevented, the waters were to be let in.

With this special exception, and the general exceptions which follow, during the term embraced by the winter season water was to be kept flowing in the Ivrea canal to a depth of 4.6 feet, in the Cigliano to a depth of 4.3 feet, and in the Rotto to a depth of 2.6 feet, according to the official gauges in the canals respectively; and the amount thus conducted was to include the winter supply due the old grantees of water-rights as well as that at the disposal of the association.

Should it become necessary at any time during the year to execute extraordinary repairs to either of the canals, or the headworks in the river supplying them, the association was obliged to suffer the loss of water for the time, without claim to rebate on its payment for the season.

Should the supply of water in the river at any time fall short of the amount sufficient to supply the flow demanded, the deficiency was to be divided between the canals in proportion to their summer volumes of flow, and be borne by the association without recourse for loss.

The state is restrained from issuing privileges to any other person or association, for the diversion of water from the Dora Baltea or Po, for irrigation in either of the three districts included within the area to be served by this association.

The state reserves, however, the right of using the three principal canals, or either of them or the secondary branches, for the irrigation of other districts, provided this can be done without diminishing the supply required by the association.

The state also reserves the right of collecting the surplus drainage waters from irrigation below the lands of the members of the association, and of re-disposing of them for irrigation in the lower district of the Lomellina.

The water is leased to the association for the use of its members, and not for sale to persons not members, and except in cases of urgent demand when a casual watering may be given the lands of some outsider, the association is prohibited from taking more water than its members require, and from delivering it to others than members for use.

In the event of the government authorizing or causing a canal to be built from the river Po, so as to command the districts to which the agreement related, the association was to have the privilege of taking the waters of the Po, in place of those of the Dora Baltea, at an advanced rate of payment, as elsewhere spoken of.*

MANAGEMENT OF THE WATERS.

The management of the diversion of the waters from the rivers, and of the upper portions of the canals down to the point of gauging in each, was to be wholly in charge of the government agents; the gauging was to be in charge of the government engineers, and thenceforward in the canals the association was to take charge of the flow of the waters and their distribution, except as might be necessary in effecting repairs to works, as will hereafter be seen.

The unit of measurement in the distribution as well as the original delivery of the waters, was to be the *module* defined and legalized by article 643 of the (Sardinian) civil code, and equivalent to 2.047 cubic feet per second.

The association was to apportion and deliver to the holders of the old water-rights the quantities of water due each, to the aggregate volume of 387 modules during summer, as they had been delivered in the past, and according to a schedule annexed to the agreement. Should the association fail to deliver these waters, the state reserved the right to take charge of the works, and deliver them.

MANAGEMENT AND MAINTENANCE OF WORKS.

The state, at its own expense, was to preserve and protect the three principal canals and all structures immediately connected with them.

---

* Such a canal was afterwards built, as will be shown in the next chapter.

The works of ordinary maintenance and repairs, on all irrigation works, was to be performed by the agents and engineers of the state, yearly, at the expense of the association.

The state assumed the responsibility of managing the main works only to an extent sufficient to deliver the waters at the points of main distribution.

Beyond that, the association was to conduct the operations and bear the expense of every kind necessary for delivering, distributing, and employing the waters leased, and also the expense of maintaining proper drainage facilities to save the surplus waters after use.

WATER-POWER AND MILLS.

The state also accorded the association the use of the royal establishment and mills of Salasco, to be used by it in the administration of the works, and in cleaning and handling the grain, rice, and other produce paid by the irrigators for water rents.

The society was to keep the establishment in repair, and be responsible for loss by fire or otherwise.

The water-power of the canals, beyond that necessary for the above named establishment, and beyond that necessary for the existing or new mills of the members of the association, devoted for their own private use, was reserved to the state.

[The introduction of new machinery into the Salasco establishment by the state, and its use by the association, was also the subject of articles of agreement, but these it is unnecessary to summarize here.]

REVENUE AND RENTS.

The annual dues from the holders of old water-rights were to remain as before, payable to the state.

All other income from the use of the waters delivered to the association, except that which might be derived by the state from the use or disposal of water power, as elsewhere explained, was to be to the benefit of the association.

The association was accorded full power to fix its rates for water to be delivered to its members as consumers.

In return for the use of the waters, the association was to pay the state for the water used by it during the summer season, at the rate of about $80 per cubic foot per second. Payment to be made before the thirty-first of December of each year, and to be collected the same as taxes.

GENERAL CONDITIONS.

All the waters, rights, etc., were turned over to the society as they existed, the society assuming all responsibility in their management, except as stipulated to the contrary in the agreement; and only in the case of their proving profitless by reason of absolute failure of water, or a raging plague, or a war being waged in the locality, could the association have recourse against the state for a remission of its dues.

The state by its engineers was to prepare and cause to be published at the joint cost of itself and the province of Vercelli, a hydrographic map in detail of the whole region covered by the agreement.

The state reserved the power to appoint a special commissioner to represent it in the councils of the association, and to care for its interests generally under the agreement. This commissioner was not to have any vote in the assembly or syndicate, but was simply to have a voice to speak for the state when necessary.

The association was required to deposit a money bond equivalent to about $60,000, to be held by the state as a guarantee of the good faith of the association and the payment of its dues.

But for the first year the association was to be permitted to use this sum for working expenses if necessary; or in after years in case of great necessity, of which the state administration was to be the judge, and upon an agreement made at the time to return it within a limited period.

[There were other provisions about the extension of the works by the state, and the use of such new works by the association, which it is unnecessary to summarize here.]

SECTION III.

ORGANIZATION OF IRRIGATION ASSOCIATIONS.

THE PRESENT LAW FOR ALL ITALY.

The Sardinian code did not contain any provision for the formation and management of irrigation associations. A law similar to that of the former kingdom of Italy, promulgated at the order of Napoleon, and hereinbefore transcribed as the law of Lombardy, made such provision in detail for Piedmont, and, as we have seen, the great irrigation association of the country was recognized by a special law sanctioning the lease of the waters of the crown canals to be used under its management.

When, in 1865, the code for all Italy was formed, however, there was incorporated into it a number of articles which declared the liberty and power of forming associations specially for such purposes as irrigation and drainage, and formulated the principles to be observed in the management of their affairs. (See, Appendix II.)

These are to be found in articles 657 to 660, inclusive, under the title, "In what way servitudes are to be exercised," and in articles 673 to 684, inclusive, under the title, "Of community property." There results from these rulings the following application for our case in hand:

VOLUNTARY ASSOCIATION OF LANDHOLDERS.

Where in a natural irrigation, reclamation, or drainage district, a community of interest exists such as to clearly render coöperation amongst the landholders advisable, in order to effect the desired purpose of irrigation, reclamation, or drainage, as the case may be, these owners may form an association to jointly act in the matter. For such free association it is only necessary that the assent of the members be had in writing, and that the by-laws under which they propose to operate be similarly recorded. (Art. 657.)

Such organizations are governed by the action of a majority of their members, who represent at the same time a majority interest in the common property and benefits of the association. Thus, to constitute a majority there must be, not only more than half of the parties at interest, but these parties must represent more than half of the total interest merged in the association itself. (Arts. 658 and 678.)

The resolutions and determinations of such a majority are binding upon a dissenting minority in the association. Assessments may be levied, which become a lien on all the property represented in the association and collectible the same as taxes. (Art. 678.)

If at any time a majority, as already defined, cannot be had for or against a proposed measure, or if the determination of a majority may threaten detriment to the interests at stake, the judicial authority of the province may look into the matter, and, if necessary, appoint some one to administer the affairs of the association. (Art. 678.)

No one can be compelled to remain a member of such an association; but lands which have been entered as represented cannot be withdrawn during the period for which they are entered, up to a limit of ten years. Thus, any person joining such an association is admitted as the representative of his certain specified property in the district, and thereupon is entitled to representation as an individual, and also as the owner of a certain interest, proportioned to the whole as may have been agreed on. He may thereafter himself withdraw; he may sell, lease, or hypothecate his interest to others, who may represent it in the association, or he may leave it without representation; but the property itself is held to the agreement for at least ten years, unless the judicial authority, on due hearing, may be shown the justice of releasing it. (Arts. 681 and 679.)

Furthermore, no individual owning a part of a property thus merged in common interest can be allowed to withdraw, even by the courts, if the part so withdrawn is such as to defeat the purpose of the association. (Art. 683.)

Each individual member of such an association may proceed against each other individual member, and compel him to contribute his share towards the proper maintenance of their common interest, unless the directory of the association releases those proceeded against, or they release themselves by an abandonment of their interest. (Art. 676.) But no individual can make any change in the property owned in common, no matter how much to the advantage of all it may appear to be, unless the others consent. (Art. 677.) The proper preservation of a common interest does not necessarily imply a change, however; so that each individual can be held to sustain the acts of every other individual associate, wherein it can be shown that such acts are necessary for the preservation of their joint interest.

Such an association can only be dissolved at the end of the time for which it was formed, or by resolution of a majority exceeding three fourths, or when a dissolution may be effected without serious detriment to the interests involved.

In the first case, when such associations are formed for definite periods of time, the fact of formation binds all the property for that time; and, hence, until its expiration, except in the cases which follow, the association cannot be dissolved.

In the second case, the majority required is not only more than three fourths the individuals of the association, but, also, a representation of more than three fourths the interest involved.

In the third case, the question—as to serious detriment to interests involved—is always to be decided by the courts; so that, an association being formed, to dissolve it before the time an application has to be made by parties interested, and the courts have to be shown that no interests are to suffer by the dissolution, before it is authorized. (Art. 660.)

COMPULSORY FORMATION OF ASSOCIATIONS.

Such is the general idea of a free association, where all the parties voluntarily join in the movement.

But the law does not stop here: We find in article 659 that the "judicial authority"—that is, the judges of the superior court of the province (an appeal being open to a higher tribunal)—"when it is a case of the exercise, the preservation, or the defense of a common right, of which it is impossible to make a division without serious injury," may order the formation of an association of all owners of lands in any such district, when a majority of them shall have demanded such organization, and the others shall not, on being heard, have been able to show good reason why the action should not be taken.

Now, "a case of the exercise, the preservation, or the defense of a common right, of which it is impossible to make a division without serious injury," may be presented in an irrigation district, and it is almost always presented in a drainage or reclamation district. These points are judged of by the courts, and, as I have in several places heretofore explained, the courts are guided in their decisions on such matters by the opinions of engineering experts, limited in number under the orders of the courts, nominated by agreement, if possible, amongst the parties interested, or, in default of such agreement, then appointed by the court itself.

Here, then, we have a result: Where a community of interest exists in a district of country such that in order to manage its irrigation or effect its reclamation or protection from floods, it is necessary, in justice to each owner and each parcel of property, that all combine in an association for the common protection or the exercise of a common right, if a majority of the owners of the property, representing a major part of the property owned, make application to the proper tribunal to have an order issued for the purpose, and, after a due hearing of all parties, it appears to the court that the interests should be combined, an order will be issued compelling such combination. The minority can be forced into the association; and, being in, as we have seen, the minority are subject to the rulings of the majority; but in the cases where associations are thus formed under orders of a court, the resolutions of the majority are subject to revision by the court. (Art. 659.)

Here we have a result parallel to that which we have before noted as an established principle in France. The principle of compulsory action where an interdependence of interest clearly demands it in order that a common good may be subserved.

As has been said, reclamation districts almost always present such cases. The French apply the rule only in cases where "the water is an enemy." The Italian law is somewhat broader in its wording, so that a proposed irrigation district, even, might present the conditions which would warrant the compulsory association of the owners of its lands, on the petition of a proper majority of them.

AUTHORITIES FOR CHAPTER XIV.

In the preparation of this chapter I have consulted and compared the following named authorities:

Moncrieff.—[Work cited as an authority for Chapter VII.] See, Chap. XV, and appendix D.

Smith.—[Work cited as an authority for Chapter IX.] See. Vol. II, P. 4, Ch. 1, Sec. 5; Ch. II, Sec. 5, and elsewhere.

Marsh.—"The evils, remedies and compensations of irrigation;" by Geo. P. Marsh (U. S. Minister to Italy). See, Report, Department of Agriculture, 1874, pp. 362–381.

Italian Code.—[Work cited as an authority for Chapter X.] See, Articles 657 to 660 and 673 to 684.

CHAPTER XV—ITALY[7];

IRRIGATION ENTERPRISE.

SECTION I.—*Forms of Enterprise—Examples of Canal Construction.*
  The Association Principle not Applied.
  Ancient and Modern Enterprises.
  The Great Modern Work—Cavour Canal.
  Organization—Management—Failure of the Company.

SECTION II.—*Concessions to Capitalized Companies—Cavour Canal.*
  The Cavour Canal Concession.
  Obligations of the Company.
  Condition of the Concession.
  Privileges to the Company.
  Benefits to the Company.

SECTION III.—*Governmental Policy and Encouragement.*
  General Policy as to Public Works.
  Prize Competition in Irrigation Practice.
  Hydrographic Survey of Italy.

SECTION I.

FORMS AND EXAMPLES OF CANAL ENTERPRISE.

THE ASSOCIATION PRINCIPLE NOT APPLIED.*

Although, as I have endeavored to show in the chapter preceding, the principle of association has been very fully developed and applied in northern Italy, in the matter of organization for the use of water in irrigation, it has not been thus availed of for the ordering of enterprise in canal works. The property owners, both poor and rich—the peasantry and gentry, and even the nobility—have participated in the formation, and now maintain the organization of associations, almost all over irrigated Lombardy and Piedmont, having for their object the leasing in bulk, and distribution among their members, the waters of the canals. But the canals themselves are almost without exception the property of the State, of municipalities, of ecclesiastical bodies, of corporations, of (in each instance) a few rich and powerful noble families or wealthy landholders.

---

* See, De Buffon, generally; acknowledge, also, letters from Hon. Geo. P. Marsh.

The association principle, of which I speak, in northern Italy, was born of and derived its strength from the necessities of the people for protection: on the one hand, protection from the floods of water which, spreading out from the great rivers, have so frequently devasted the lower parts of the fair valley of the Po; and on the other hand, protection from water monopoly which long ago became implanted on the higher lands, and spread its blighting influence even more widely than did the floods theirs.

It was early realized that association, unification of effort—if not voluntary, then by compulsion—was absolutely necessary in the low lying districts, in order to compass a respectably efficient resistance to the spreading of the army of flood waters which the Alps and the Apennines periodically sent forward towards the sea. The march of these floods was just as much an invasion of the country as was the advance through the fields of any of the armed and ruthless hordes of men, which northern and middle Europe have so often in historic times sent trooping over the plains of the basin of the Po.

As organization was necessary to resist the human flood (but, alas, Italy, because disorganized, has not always effectively interposed such resistance), so organization on a broad scale was necessary to control the march of waters, in this same country. That the object of this organization has not always been accomplished is a matter which need not concern us here. The facts are recognized, that many thousands of acres of a most productive country have been cultivated for centuries, under the protection of works controlled and maintained by district organizations of the people, under government supervision, that the country could not have been inhabited otherwise than by a successful defensive war against the invading armies of waters, and that experience shows the general scheme of organization, and operation, so far as the political problem goes, to be the best ever adapted to measurably free people; and we can well understand its increased degree of adaptability under social and political conditions grounded on an advanced position of freedom and intelligence of the masses.

Commencing, probably, in this necessity for protection against floods, associations of landholders in districts were formed for pro-

As a natural consequence of this turn in events the proprietors of the canals have themselves organized in the several districts, and great clashings and conflicts of interest have resulted; but when opposing interests are thus locally organized they are easier to deal with by third parties, and so the government is the mediator in this instance, and through these organizations, formed under general laws, it exercises that supervision which now keeps comparative quietude and admits of a corresponding degree of prosperity.

CANAL WORKS AND ENTERPRISES.*

With very few important exceptions, the main canal works of northern Italy were built so long ago and under political and social conditions so different from those which are present in our country, and even in Italy at this day, that we find but little of a positive nature, in the forms of enterprise under which they were carried out, by the study of which we may profit. There was, at the dates of early works, no legislation with respect to irrigation enterprise, and no administration worthy of the name.

The following notes will convey an idea of the origin of the great works of the country:

Lombardy.—Eleventh century—Ancient works of the Romans in and about Milan, restored and extended.

Twelfth century—Further extension of these works. The monks of Chiaraville obtained rights to the waters of the Vettabbia, and utilized them. Construction of the great canal of the Ticino; of the canal of Battaglia, and of Reno, and of many others. This was a period of great activity in the construction of works.

Thirteenth century—The great canal—Naviglio Grande—even yet the largest in Lombardy, was completed. The canal Muzza and other great works carried out.

Fourteenth century—The Naviglio Cavico, deriving its supply from the Oglio, constructed. The great canal from Pavia to Milan built.

Fifteenth century—The canal Martizana commenced and finished within a few years—an exceptional case. The canal Bregnardo also promptly carried out in a reasonably short time.

Sixteenth century—The great lines of irrigation were extended, and many minor and secondary works built. The canal of Paderno, the only main work of importance, commenced; but it was not finished until late in the eighteenth century.

Seventeenth century—Dominion of Spain over the country from

---

* See, De Buffon, Vol. I., Ch. I, and, also, pp. 123, 134, 153, 162, 180, 196, 227; also, Smith, Vol. I., pp. 100-102, and 196-202.

the last century. Great activity in irrigation works of detail, and systemization of practice; but no new large works.

Eighteenth century, and to the middle of the nineteenth—No new large works of irrigation in Lombardy, except the new canal of Pavia, executed in the time of the former kingdom of Italy, in the early part of this century, by order of Napoleon.

Piedmont.—Fourteenth century—Most ancient existing canals built. Early part of century: The Roggia or Gattinara canal built. The canal Busca and Santirana, from the Sesia, and the canal Langosco, from the Ticino, built in the latter part of the century.

Fifteenth century—The canals Rotto and Dorea, from the Dora Baltea, built. Also, the canals Commune of Gattinara, Mora, and Sforzesca, had their origin in this period.

Sixteenth century—The only important work due to this century, in Piedmont, is the canal Coluso.

Seventeenth century—The canal Ivrea restored, after having been destroyed.

Eighteenth century—The canal of Cigliano, with its branches, was constructed.

Nineteenth century (early part)—The canal of Charles Albert constructed.

With one important exception, which is to be written of in the next paragraphs, the canals above named constitute almost all of the important works in the great irrigated region east of the Po, for although there are very many secondary and branch canals, and an immense number of distributaries, the main works are not numerous but large. The topography of the country has not admitted of cheap works, and the early policy of the government did not encourage opposition.

THE GREAT MODERN WORK—CAVOUR CANAL.*

Remembering what has been said in the introduction to the ninth chapter of this report, about the form and size of the valley of the Po and the distribution of its water-ways, we are prepared to see at once the bearing of the Cavour canal problem.

There had been a number of canals brought from the rivers which enter the valley of the Po from the Alps and course across the great plain of Piedmont to the main stream running easterly at its foot, but previous to 1844 the idea of calling upon the Po itself to contribute a portion of its waters to this field of irrigation industry, it appears, had not been seriously entertained. The probable great cost

---

* See, Moncrieff, Ch. XIV.

of the work had deterred even an examination of the project, so it was not fully realized until about the date mentioned, that the scheme was even feasable.

A canal on this route would have to cut across the natural drainage lines and, also, other canals of the country, and these were so formidable as obstacles to a great artificial water-way, that it probably appeared to be an undertaking beyond reach. It was as though a canal as large as the six largest in this state combined in one, was proposed to be constructed from Red Bluff on the Sacramento river, around the eastern margin of the Sacramento valley, crossing the Feather, Yuba, and Bear rivers, and the intervening creeks magnified into torrents, and also half a dozen or more other large canals and any number of medium sized and small ones. This was about the aspect of the Cavour canal project. The water-way was to give passage to 3,885 cubic feet of water per second, and diminishing in capacity at successive main points of distribution, it was to extend from the Po at Chivasa, at the head of the main valley, as it were, around the north side of the valley to the Ticino, one of the principal rivers entering the plain from the north.

When in 1854 the project had been quite thoroughly examined and estimated upon, it was the intention of the Sardinian government to carry it out as a public work, seeing that a large part of the country which it would command was already partly supplied by government canals, yet the supply of water was short and a great demand existed for an additional amount. But the Crimean war crippled the resources of the government, so that when a company of English capitalists proposed to take the matter in hand, the government assented, and, hence, resulted the concession to the Company General of Italian Irrigation in 1862, of which an abstract will be given in the next section of this chapter.

The principal object of the work was to supply water to the great canals already existing, particularly at the season of low stage of the rivers from which they drew their supply, at which season the Po had a surplus of water to spare.

The originally estimated cost was $7,070,000. The contract for the canal was let for $8,875,000. Contingent expenses, damages, and other matters raised the total estimated cost to $10,660,000. The purchase of the crown canals—a part of the agreement, as will hereafter be seen—and expenses accruing before income was realized, brought the total sum to be paid out by the company up to $16,600,000. The nominal capital of the company was $16,000,000, but although the government guaranteed the interest at 6 per cent on that amount, from the time the work was opened, the stock of the company never went nearly to par, its actual resources never exceeded $12,200,000, so that it failed, and the works were thrown back on the hands of the government.

Nevertheless, and although the project had so sad an outcome for the stockholders, the canal was built, and is a most noble work, and the original transaction between the government and the company affords an instructive example of government policy towards irrigation enterprise.

SECTION II.

CONCESSIONS TO CAPITALIZED COMPANIES.

THE CAVOUR CANAL CONCESSION.*

The history of irrigation in Italy does not afford many examples of concessions for purposes of irrigation in such form that they can be studied in details from the standpoint of the political economist. The general ideas of the policy of the Italian governments have been already outlined in this report, but there is no such stock of practical examples to be drawn from for illustration in detail, as we have found in France. Nevertheless, there have been some concessions of late date worthy of mention and analysis, but with the exception of that to the Company General of Italian Irrigation, for the construction of the Cavour canal, there is no data at hand in a form which enables me to make use of it in the necessarily hurried preparation of this report.

This Cavour canal concession was embodied in an agreement between the government ministers of finance and of agriculture, and six individuals standing for the company, and in a law bearing date of August 25, 1862, sanctioning this agreement.

I have made an analysis of these documents, and grouped their principal points under headings as follows: (1) Obligations of the company; (2) Conditions of the concession; (3) Privileges to the company; (4) Benefits to the company; and in this form the matter is now presented.

An Analysis of the Concession for the "Company General of Italian Irrigation" to Construct the Cavour Canal—1862.

(1) OBLIGATIONS OF THE COMPANY.

The grantees became obligated as follows:

(1) To form a company for the construction and working of a canal by which should be diverted constantly from the river Po a quantity

---
* See, Moncrieff, Appendix C.

of water not less than 110 cubic metres (3,885.2 cubic feet) per second (supposing such a discharge to exist in the river).

(2) To combine the waters of said river with those of the Dora Baltea, for the irrigation of the Novarese, Lomellino, and Vercellese districts, in accordance with the law of third July, 1853.

(3) To comply with the project of the government engineers in every respect.

(4) To have the headquarters of the company at Turin.

(5) To organize within two months from the promulgation of the law approving of the agreement.

(6) To submit the regulations of the company to the government within a month from the promulgation of the law.

(7) To construct, entirely at its own expense, the said canals, with all the works belonging to, in connection with, or dependent on them, for taking into and passing along the canals the constant discharge of water mentioned above.

(8) To commence the works within six months from the promulgation of the law.

(9) To complete the canals in every way, within four years from the commencement of the works, providing for every occurrence, and preparing for every event, ordinary or extraordinary, even of the greatest influence, without having power to exempt themselves from the liabilities assumed, and without having any claims to compensation or indemnity.

(10) To observe the contracts made by the government with the Association General of irrigation to the west of the Sesia, and those which are in force with other parties, and to satisfy the burdens, cares, responsibilities, liabilities, and obligations belonging to the said canals and property, the state considering itself relieved from every species of annoyance that may arise therefrom.

(11) To respect existing grants of motive power for the service of industrial establishments.

(12) To carry out, at the request of the government, the construction of catch-water and branch canals, even as far as beyond the right bank of the Po, near Casale, on the basis and guarantee, and with the advantages agreed on for the principal work.

(13) To obtain possession of canals, springs, water-courses, and portions of water, in the same manner and under the same terms.

(14) To raise a capital for the execution of the works, of eighty millions of *liras* ($16,000,000), of which fifty-three million four hundred thousand ($10,660,000) are reserved as a fixed capital for the construction of the new canal, inclusive of interest during the con-

struction; twenty millions three hundred thousand ($4,060,000) shall be laid out on the payment of the price of the grant of the crown canals derived from the Dora Baltea and Sesia, and the remaining six millions three hundred thousand *liras* ($1,260,000) on the purchase of canals and volumes of water of private property, and on the formation of other canals.

(15) To submit for the approval of the government the projects of all the new works contemplated in the grant.

(16) To execute the supplementary works at its own expense, which the government deems necessary to ensure the constant supply of the main canal, and also to pay all expenses in connection with the government inspection, superintendence, and approval of the works.

(17) To be responsible for the preservation of the effects included in the grant, with all things pertaining thereto, in the manner and terms laid down in the list.

(18) To hand over to government all of the above mentioned effects at the end of the grant, in a proper and fair state of repair.

(19) To gauge the waters of the canal to be derived from the Po, and carried beyond the Sesia, above the head of the first outlet of said waters, by means of a hydrometer, made according to the best hydraulic rules, and referred to bench marks, in order to give a discharge of not less than ninety cubic metres (3,178 cubic feet) per second, except when there is a deficiency in the waters of the Po, in which case, the company shall make up the difference with the waters of the Dora Baltea.

(20) To lease out when called upon, to a general association of proprietors west of the Sesia, all the water which flows past the gauge above mentioned, at a price to be determined on by the government in concert with the society.

(21) To supply with the waters which are not thus leased out, the parishes, small associations and proprietors, at a price fixed by government.

(22) To retain in its service on the crown canals, of which it shall be given the use, at whatever salary the government shall establish, those officials employed on the direction and care of the said canals, who shall be specified in a list, and also to pay the annual salary of those on the reserve or retirement list, in terms of the laws in force in such matters.

(23) To provide the volume of water necessary for the irrigation of that piece of land in the Lombardian territory lying above the Grand canal of Milan, to its left, provided the government sees fit to prolong the canal beyond the Ticino.

(24) To pay to the widow and descendants of the late surveyor, Francesco Rossi, who first pointed out the possibility of utilizing the waters of the river Po, for the Vercellese and Lomellino territories, the reward that was promised to him while alive, namely, the sum of 50,000 *liras* ($10,000), in the manner and terms which shall be fixed by government.

(25) To deposit in the state treasury as a guarantee, within fifteen days from the day of the publication of the law ratifying the grant, a million of *liras* ($200,000) in paper of the Italian national debt, at the par value; this deposit to remain until there shall have been executed works for the construction of the canal, to a value of ten millions of *liras* ($2,000,000).

(26) To observe all the conditions and securities necessary to develop and harmonize the essential terms of the grant, and to guarantee as far as possible the reciprocal interests of the state and the company.

(2) CONDITIONS OF THE CONCESSION.

(1) That if the company uses the royal canals which the government grants them, it must pay for the same canals and property 20,300,000 *liras* ($4,060,000), to be paid to the treasury in three equal portions, within twelve months of the promulgation of the law, by means of bills on banks approved of by government, payable at six, nine, and twelve months, which may be discounted on the exchange of London.

The payment of the said bills should be made to the treasury immediately upon the promulgation of the law.

(2) That at the end of the said fifty years, the whole property and free disposal of the canal shall fall into the possession of the state, without any sort of compensation being due to the company.

(3) That the additional works carried out by the company at the request of the government, and the contracts for purchase, be approved of by law.

(4) That the expenditure on the formation of new canals, besides the main one, shall be fixed by general consent, or by means of arbitration, and that the cost of purchasing them shall be according to what is agreed on with the sellers.

(5) That the company accepts as definite the sum of 53,400,000 *liras* ($10,660,000), as an estimated cost, and assumes in consequence, entirely at its own risk and peril, whatever expenses there may occur in excess on the construction of the works necessary to ensure the constant supply and the constant passage of the volume of water stated in article

1, excepting the provisions with regard to the cost of maintenance and repairs.

(6) That the coupons of the bonds issued by the company shall be countersigned by a government commissioner. The sum raised by the bonds shall be deposited in the public treasury, to be issued to the company according to the actual requirements of the undertaking.

(7) That the company provide in due time the necessary sum on which the government guarantees the interest, and that it pays to the said treasury a commission of 2 per 1,000.

(8) That the bank in London through which the government pays the interest shall give notice of, fifteen days before they fall due, the coupons or bills which may have been presented for payment.

(9) That the government has the right of superintending the execution of the works, and of approving them before they are carried out.

(10) That the government has the right, within four years from the commencement of the work, of prescribing all the supplementary works which may be necessary to ensure the constant supply of the main canal.

(11) That the government has the right of watching over the proper execution of whatever forms a part of the present concession; as, also, of inspecting the management of the company in its financial affairs.

(12) That stock shall be taken by the government commissioners, in contradistinction to the company, of all the effects included in this grant, immediately after the company have undertaken the execution of it, in order to establish an efficient control over them.

(13) That the amount of water-rate and the price of water-power, except where otherwise specified by government, shall be fixed by agreement between the company and the government, and that the price must not be varied without consent of the government.

(14) That the final alienation of the water which the company has the right to carry across the Sesia must be approved of by law; and that in this case the profits of the sale shall be deducted from the capital of the company, and the state shall pay it the interest agreed on for the rest of the capital.

(15) That the obligation of the government guarantee is only conditional, and shall only take effect when the net income does not amount altogether to the sum necessary to make good the guaranteed interest and refund. (The net income consists of the revenue of every description, including the rents and the returns of the canals and of the property handed over by the state, deducting all the charges for maintenance and repairs, both ordinary and extraordinary, besides those for administration.)

(16) That government reserves to itself the power of prolonging the new canal beyond the Ticino, to benefit that portion, hitherto unirrigated, of the Lombard territory lying above the grand canal of Milan to its left, giving the preference of the grant of it to the present company on equal conditions.

(17) That all questions arising between the company and the government, on the meaning and execution of the present contract, the decision shall be referred to three arbitrators—the one chosen by the company, the other by government, and a third by the president of the court of appeal sitting in Turin. The decision, provided it does not exceed the limits agreed to by the contending parties, shall be final and obligatory.

(18) That after twenty years of the occupation have transpired, it shall be in the power of the state to redeem the grant, paying to the company the capital corresponding to the mean net annual income of the last three years, at the rate of five per cent, with a deduction of the sum already refunded by the guarantee paid by government.

(19) That the general approval of the plans are to be given by government, within the year of the commencement of the canal.

(21) That the agreement be strictly limited to the expenditure of bare capital of 80,000,000 of *liras*, and that it have its full effect only when the sum in excess of the two capitals of 53,400,000 *liras*, and 20,300,000 *liras*, is being advantageously laid out on the works, and on the purchase of those works mentioned before, as being supplementary to the canal.

(3) PRIVILEGES TO THE COMPANY.

On the foregoing conditions, the company shall have privileges:

(1) To introduce from abroad all materials necessary for the construction and maintenance of the canal, with a reduction of 50 per cent on the customs duties, and to introduce free of customs duties those instruments and implements of work which the company may need to carry out the various operations of the canal, under compliance with the conditions, which, for the security of financial interest, may be established by the minister.

(2) To be exempt from all registration duties on deeds and contracts, arising from an execution of the grant, and subject only to the fixed duty of one *lira*.

(3) To use the royal canals derived from the Dora Baltea, with the branches of the same, and everything connected with, or depending on them, including the factories, mills, thrashing mills, and every other workshop there belonging to the state.

(4) To enjoy the use of the said state canals from January 1st,

1883, up to the end of the grant, and after that date, the state shall resume full and free disposal of the same.

(5) To enjoy the use of the new canal to be constructed for fifty consecutive irrigating years, beginning from the year in which the newly constructed canal shall commence working, if opened before the middle of April.

(6) To raise the capital required for the execution of the grant, partly by means of shares for the fixed sum of 25,000,000 *liras*, and partly by bonds bearing interest at six per cent, to the amount of 55,000,000 *liras*.

(7) To take the place of the state in carrying out the objects of the grant, and to insist on the observance of all rules in force.

(8) To alienate, with the consent of the government, all or part of the waters carried beyond the Sesia.

(9) To recover all rents of every kind due the company, in the same way, and with the same privileges as the law directs for the public taxes, by the appointed collector.

(4) BENEFITS TO THE COMPANY.

(1) All works in connection with the canal are declared of public utility.

(2) The profits of the new canals, besides the main one, shall belong exclusively to the company for the whole period of the concession.

(3) On the cost of construction of the canal, and on the sum raised according to agreement, government guarantees to the company:

(*a*) An annual interest of 6 per cent to be paid only for the objects of the grant, from the day in which the fifty years begin to be counted.

(*b*) A refund of .3444 lira per cent on the sum expended on the canals to be derived from the Po, and on the royal canals derived from the Dora Baltea and Sesia, and on the other items of the balance of the capital, a refund in proportion to the number of years not yet elapsed, of the grant.

(4) Government engages to prohibit the opening of new *fontanili* (springs) along the projected canals for a distance of 300 metres from the main canal, 200 metres on the principal supply canals, and of 100 metres on the main branches taken off the said canals by the concessionary company.

(6) Government engages to provide that the communes, provinces, and responsible bodies be authorized to take that number of shares and bonds of the company that they may see fit, contracting loans to meet the payment of said shares and bonds, and mortgaging their

income for three years ahead for the payment of the interests, and for the repayment of the capital, if it should necessarily exceed the natural limits of their special taxes.

SECTION III.

GOVERNMENT POLICY AND ENCOURAGEMENT.

GENERAL POLICY AS TO PUBLIC WORKS.*

The Italian government, although of late years advancing rapidly in the scale of enterprise, and upon a line of policy looking directly to the development of the agricultural resources of its territory, has not to this time gone nearly so far as has that of France in the way of encouraging irrigation enterprise.

Italy owns more great works of irrigation than does France. But the present government has fallen heir to them from the governments and other constructors of long ago. Where irrigation is most demanded in Italy works were already constructed when the present government came into power a few years ago.

Other great interests were demanding attention, so that the improvement of rivers and construction of great drainage works has been more in the line of government effort of late years than has the extension of irrigation facilities.

PRIZE COMPETITION IN IRRIGATION PRACTICE.§

One step made quite recently is worthy of special mention. It will be remembered, from a reading of a former chapter,† that, in 1874, the French government, by decree, offered prizes for the best examples of irrigation practice, as an encouragement for the economical use of waters, and a means of acquiring information about irrigation, to spread abroad amongst its agriculturists.

The Italian government followed closely in this line of policy, and by a decree issued in 1879, offered prizes not only for the best examples of irrigation practice, but also for the best examples of agricultural drainage, of colmatage,‡ and of drainage and irrigation combined. The following is the full text of the decree:

---

* Letters from Hon. Geo. P. Marsh.
§ Documents from Hon. George P. Marsh.
† See, pp. 145–148, *ante*.
‡ *Colmatage;* see, foot-note, p. 82, *ante*.

Royal Decree, which opens a prize competition for works of Drainage, of Irrigation, and of Drainage and Irrigation, combined. June 19, 1879.

HUMBERT I,

By the grace of GOD and the good will of the NATION,
KING of ITALY.

In accordance with the resolution of the council of agriculture, at its session of 1879, which provides for the arranging of a prize competition for works of drainage, of irrigation, and of colmatage;
Conforming to the proposal of our minister of agriculture,
We have decreed, and do decree:

ART. 1. There is opened a competition, with the following prizes:
Two of 4,000 *lire* ($800) and a gold medal; two of 3,000 *lire*, one with a silver medal; and three of 2,500 *lire* and bronze medals, or a work of art of equal value, in favor of a private individual, or an association that executes in the interest of agriculture, and with good results, creditable works of:
(a) Drainage.
(b) Irrigation.
(c) Drainage and irrigation combined, using for irrigation the drainage water collected.
(d) Colmatage, alternating with cultivation.

ART. 2. Drainage, sub-letter (a) of the preceding article, must embrace an area of marshy land not less than fifteen *ettari*.
Irrigation, sub-letter (b), an area not less than twenty *ettari*.
Drainage and irrigation combined, sub-letter (c), an area not less than thirty *ettari*.
Colmatage, sub-letter (d), an area not less than ten *ettari*.

ART. 3. Drainage may be accomplished with open ditches or any system of drain-pipes, but must be so complete as to make the land well cultivable for winter wheat.

ART. 4. Irrigation must be done according to rule, abundant distributing ditches must be provided, so that water may percolate without too great resistance.

ART. 5. Water derived from drainage works may be conducted for irrigation, to lands at a considerable distance, but it must be done in a regular canal which will not obstruct its flow.

ART. 6. Crops irrigated may be diversified to suit the character of the lands.

ART. 7. The explanations of the works entering into competition must be transmitted to the minister of agriculture, industry, and commerce no later than March 30, 1880. The work must not have been commenced before the present date, and it must be competitive work, excepting colmatage in progress, of which the following article (8) treats.

ART. 8. The work, sub-letters (a), (b), or (c), must be completed no latter than March 31, 1882.
Those, sub-letter (d), are divided into two classes:
(1) Colmatage in progress, by means of which (the colmatage itself having been executed with good result) for two years at least, preceding the time specified in article 7, a crop has been raised each season after the drying.
(2) Colmatage not commenced at the time of the publication of the competition, but regularly carried on with satisfactory result to the date specified in the preceding paragraph of this article.

ART. 9. The minister of agriculture, having received the statements of the work to be entered in competition, will have the condition of the land examined.

ART. 10. The work finished in accordance with article 8, the minister himself shall order another examination to ascertain whether the competitor has satisfied the conditions of the competition.

ART. 11. The results of the competition shall be presented in proper form to the council of agriculture, which has power to award the prizes.

ART. 12. I order, that this decree, provided with the seal of state, be inserted in the official collection of laws and ordinances of the kingdom of Italy, and command that every one interested observe it and cause it to be observed.

Dated at Rome, June 19, 1879.

HUMBERT.

This action is one in the interest of the individual cultivators, and shows the Italian government to be alive to the importance of the agricultural development of the country, and to realize the part which irrigation, drainage, and colmatage must play in such work. The object is, of course, in so far as irrigation is concerned, to incite irrigators to thoroughness and system in the preparation of their lands, and to care and economy in the use of waters, that it may become known what can be effected by such means, and thus not only new irrigations be encouraged but old ones remodeled and the better managed.

HYDROGRAPHIC SURVEY OF ITALY.

Finally, the Italian government has in progress probably the most thorough study of its water-courses and water supply system that has ever been attempted for any country. So that it is in a position, which is continually being bettered, to deal with its waters and streams in a business-like way. Knowing what waters there are, what claims there are against them, what use is made of those diverted to satisfy such claims, and what can be effected, the government of Italy is in a position to advance its agricultural interests, with a full understanding of the outcome of every proposed move; and to prevent by exposure of error those movements which must result only in litigation and loss.

It treats such questions as every prudent business man would those of his affairs.

www.ingramcontent.com/pod-product-compliance
Lightning Source LLC
Chambersburg PA
CBHW022104230426
43672CB00008B/1270